What Things Do

What Things Do

PHILOSOPHICAL REFLECTIONS ON TECHNOLOGY, AGENCY, AND DESIGN

PETER-PAUL VERBEEK
TRANSLATED BY ROBERT P. CREASE

The Pennsylvania State University Press
University Park, Pennsylvania

The translation of this book was made possible by a publication grant from the Netherlands Organization for Scientific Research (NWO).

Figure 2 from Bruno Latour is reproduced from W. Bijker and J. Law, eds., *Shaping Technology/Building Society* (Cambridge, Mass.: MIT Press, 1992).

Originally published as *De daadkracht der dingen: Over techniek, filosofie en vormgeving* by Boom Publishers, Amsterdam. Copyright 2000 Peter-Paul Verbeek.

Library of Congress Cataloging-in-Publication Data

Verbeek, Peter-Paul, 1970–
[De daadkracht der Dingen. English]
What things do : philosophical reflections on technology, agency, and design / Peter-Paul Verbeek ; translated by Robert P. Crease.
p. cm.
Includes bibliographical references and index.
ISBN 978-0-271-02540-7 (alk. paper)
1. Technology—Philosophy.
I. Title.

T14 .V48 2005
306.4'6—dc22
2004020480

Printed in the United States of America
Published by The Pennsylvania State University Press, University Park, PA 16802-1003

Second printing, 2005

The Pennsylvania State University Press is a member of the Association of American University Presses.

It is the policy of The Pennsylvania State University Press to use acid-free paper. Publications on uncoated stock satisfy the minimum requirements of American National Standard for Information Sciences—Permanence of Paper for Printed Library Material, ANSI Z39.48–1992.

Contents

Preface and Acknowledgments vii

Introduction: To the Things Themselves 1

Part I: Philosophy Beyond Things

1 Technology and the Self 15

2 The Thing About Technology 47

Part II: Philosophy From Things

3 Postphenomenology 99

4 A Material Hermeneutic 121

5 The Acts of Artifacts 147

6 Devices and the Good Life 173

Part III: Philosophy For Things

7 Artifacts in Design 203

Bibliography 237

Index 243

Preface and Acknowledgments

Listen more to things than to words that are said.
—Birago Diop, "Breath"

In this book, I aim to answer this call of Birago Diop literally, by making a philosophical analysis of the relations between human beings and material objects. Taking things seriously is not a self-evident thing to do, especially not in philosophy. Whoever thinks we are living in a materialist age is wrong. The Western world does not attach as much value to things as one might expect. The enormous quantity of prematurely discarded objects at our garbage dumps bears witness to this. At the level of theory and reflection as well, materiality is often neglected. Social scientists usually prefer to occupy themselves with humans, and philosophers with words and ideas. Yet in the past century all kinds of new objects, especially technological devices, have come to play a profound role in what we do, how we perceive and interpret the world, and what choices and decisions we make. To understand these activities of things, these "acts of artifacts," is the main ambition of this book.

But even a book about things cannot be written without humans who inspire and support it. At this place, I would like to thank them cordially. My colleagues of the Department of Philosophy of the University of Twente form an environment every scholar can be jealous about. I would especially like to thank Hans Achterhuis, Petran Kockelkoren, and Pieter Tijmes, for all of our engaged, congenial, and critical discussions, and H. Floris Cohen for his stimulating intellectual enthusiasm and his active support of the realization of this English translation. The Eternally Yours foundation, with the open and innovative approach to industrial design of Liesbeth Bonekamp, Ed van Hinte, and Henk Muis, provided a hospitable and fruitful basis for confronting philosophical analysis with the daily practice of product design. A special word of gratitude goes to Don Ihde, for our inspiring discussions about phenomenology and technology, and for his unremitting efforts for the coming about of this translation. To Setargew Kenaw Fantaw, I am greatly indebted for his proficient review of the final manuscript. This translation, carefully rendered by Robert P. Crease, was made possible by a publication grant from the Netherlands Organization for Scientific Research (NWO).

Enschede, the Netherlands
November 2004

Introduction: To the Things Themselves

THE DEATH OF THINGS

What role do artifacts play in our technological culture? The technological developments of the past century have made this question more pressing than ever. Our society is saturated with tools, appliances, and other assorted objects that strongly shape the course of daily life in many ways. Our personal interactions are inextricably bound up with telephones and computers; our traveling with bicycles, automobiles, trains, and airplanes; our eating with refrigerators, ovens, and microwaves; our leisure activities with televisions, videos, and electronic devices. Even our being born, staying healthy, and dying depend on a wide variety of medical instrumentation. What effect do these things and artifacts have on us? How can we understand their role in our lives?

Until recently, philosophers have paid scant attention to this question, preferring to devote their attention to words and ideas rather than material things. The history of philosophy since Plato has tended to assign a secondary status to material and changeable things in favor of eternal and unchangeable ideas. The "linguistic turn" that philosophy took in the past century is only the most recent manifestation of this *horror materiae.* Since this turn, the language in which human beings speak about reality is thought to determine what may count as reality. But this turns things into little more than projection screens for our interpretations, reduces them to the words with which we describe them, and fails to give them their due, their proper weight. However important the role of language may be, its absolutization assures that things and artifacts can no longer be philosophically analyzed, only talked around.

But not only philosophy has failed to recognize the significance of things and their materiality—so has contemporary postmodern industrial design, whose products are devised principally to serve as signs rather than material things, as symbols or icons for their owners' lifestyles. Postmodern consumers purchase objects less for their materiality than for their ability to express the kind of person they want to be taken as. "We sell lifestyles, not appliances," as Braun Electronic GmbH business executive Godehard A. Günther is quoted in a textbook for industrial designers. The same textbook quotes the French luggage manufacturer Louis Vuitton as saying: "We produce travel merchandise of the highest quality, of course . . . but apart from that we also sell myths and dreams, and a particular style of traveling. Luxury, independence, and the feeling of flexibility characterize the owner of Vuitton luggage" (Bürdek 1996, 228). Products are deliberately designed for lifestyles whose latest trends are tracked by market researchers.

Despite all the recent talk about the "material world" and "modern materialism," we have managed to expunge artifacts of their materiality both in our thinking about and in our design of them. Now that we have survived the death of God and the death of the subject, we seem to be faced with the death of the thing. Therefore, as an alternative to the prevailing Platonist and immaterialist approaches to artifacts and materiality, I shall be trying to develop a philosophical perspective in which more justice can be done to the role of artifacts in contemporary culture and in the everyday life of human beings.

The attempt to give artifacts their due is, of course, not without danger. For instance, one group of thinkers in the history of philosophy has lamented the loss of materiality by characterizing it as a form of *alienation*. These thinkers, that is, understand changes in our material world as a loss of authenticity, as a growing estrangement from what things "really" are. The linguistic turn and the rise of postmodernism have made this view untenable, however, for these movements have demonstrated the ultimate senselessness of entertaining the hope that human beings can have access to "reality itself." The contact that human beings have with reality is context-dependent: "reality" is not unequivocally reachable, let alone able to function as a norm. The challenge of a philosophy of materiality, as I see it, therefore consists of navigating between the shoals presented by a too-conservative philosophy of alienation and by a too-radical strict linguistic philosophy. Only in this manner can things be brought back from the dead in a creditable manner. An approach to things must be found that neither reduces them to non-thingly elements nor falls prey to the alienation thesis.

This study aims to develop this new way of thinking in particular with respect to technical or technological objects, which are the most distinctive class of objects in contemporary technological culture.[1] The case for a "thingly turn" is most pressing within the philosophy of technology, for it, too, has neglected the value of artifacts, with serious consequences for the standard picture of technology that it has produced. This picture is dominated by the alienation thesis described above and has a strongly negative cast. It represents technology as a radically transformative power that estranges human beings from themselves, from each other, and from reality itself. Technological culture is seen as transforming human beings into cogs in a social machine, and as transforming reality into raw material that can only be approached via domination and control.

This standard picture of technology, evident in the work of Karl Jaspers and Martin Heidegger, has been subjected to harsh critique in recent years. Empirical studies of the development and use of specific technologies point to the need for a more nuanced approach. But it would be rash to conclude that the classical philosophy of technology has become entirely obsolete. The *questions* it posed are still pressing; its problems lie in the ways it has gone about pursuing the *answers*. The first chapters of this book show, on the basis of a close analysis of Jaspers's and Heidegger's philosophies of technology, that each of these thinkers approached technology from frameworks that reduce technological artifacts to nontechnological elements such as social organization and the will to power. This analysis shows that the failure to take artifacts seriously led the classical philosophy of technology astray.

The questions posed by the classical philosophy of technology deserve a new set of answers, one that does justice to the concrete presence of technological artifacts in our culture. In this book, I attempt to do so by taking up concepts and theories provided by several contemporary philosophers of technology, including Don Ihde, Albert Borgmann, and Bruno Latour. I shall undertake a systematic analysis of the relations between human beings and technological objects, wherein I discuss the connection to contemporary industrial design, which continually turns out products that play extensive roles in shaping everyday life. The resulting "philosophy of artifacts" aims to shed more light on this process and to allow the philosophy of technology to be productively turned toward practical design issues.

1. By the term "technology" I follow current usage and generally mean to refer to the specifically modern, "science-based" technological devices of the sort that began to emerge in the last century. The term "technical" is broader, and describes all sorts of techniques and their associated objects. The "technical" thus encompasses the "technological." For stylistic considerations I shall use both interchangeably to refer to modern scientific devices.

THE THING ABOUT THE PHILOSOPHY OF TECHNOLOGY

How to approach technology in terms of "things" or artifacts is not obvious. As already mentioned, classical philosophy of technology has devoted little enough thought to the role of technological artifacts in contemporary society and daily life. It tends to approach technology in terms of what technology requires or presupposes; the kinds of technological thinking it fosters, for instance, or the principles around which our social life must be organized in order for technology to function in it. If concrete technologies enter into these analyses, it is merely as illustrations of the origins or presuppositions of technology and not as points of departure for investigation into how technologies concretely shape human life. Classical philosophy of technology has come under severe criticism for this kind of approach,[2] and in what follows I shall try to make clear that it is precisely the neglect of a treatment of concrete technological artifacts that opens it up to such critique.

Classical and Empirical Philosophy of Technology

Classical philosophers of technology have painted an excessively gloomy picture of the role of technology in contemporary culture, worrying that technology would end up alienating human beings from themselves and reality. The ever more dominant technological way of thinking, they feared, would discourage human beings from approaching reality as inherently valuable and would instead encourage them to approach it as raw material. In the technological organization of contemporary social life, human beings no longer appear as unique individuals but only as functional workers needed to keep the highly structured apparatus of mass production working. This bleak diagnosis is an understandable reaction to the rapidity and pervasiveness with which technology has changed our culture. But with the appearance of more and more empirical research into the development and use of specific technologies, the self-evidence of this diagnosis has steadily declined. In the eyes of contemporary critics, the judgments of the classical philosophers of technology were too abstract and sweeping: abstract in that they failed to connect with concrete technological practice, and sweeping in that they were couched in blanket terms of "Technology" with a capital *T*, leaving no room for different kinds of descriptions of different kinds of technologies.

2. See, for instance, Achterhuis (2001a, 1–9).

The empirical studies that scholars researching science and technology have conducted over the past few decades therefore pose a serious challenge to the philosophy of technology. These studies have called into question the classical picture of technology as an all-determining and alienating power, and have brought to light a much more complex side. By researching specific technologies in concrete applications, they have brought to light the fact that technologies have different impacts in different contexts. The supposed determinism of technology appears to be weaker than is presented in the classical picture; while technologies do indeed strongly shape the form and the context in which they function, this happens in a more differentiated and local manner than in the traditional view.

How much does one really say about the microwave oven, for instance, if only that as an artifact of "Technology" it is to be understood as reflecting the technological will to power or the functionalist organization of social life? Consider, for instance, the detailed analysis that Cynthia Cockburn and Susan Ormrod have provided of this technology in their study *Gender and Technology in the Making.* As with many new technologies, the microwave oven was initially a "gizmo," a high-tech toy marketed primarily at men. It was sold in electronics stores alongside video recorders and stereo merchandise. Once this technologically sophisticated context market was saturated, the microwave then made a complete turnabout and was marketed more and more as an ordinary cooking appliance—no longer sold in electronics stores among the "toys for the boys," but as a regular household item. The sale of microwaves then moved to kitchen appliance stores, found amongst the refrigerators and the ovens.

This transition, too, was accompanied by the awareness that microwave ovens had to be marketed to women, which brought about a remarkable change in their design: all at once their operation was vastly simplified (Cockburn and Ormrod 1993, 80–91). Because women were considered to be not terribly technologically sophisticated, the operating "bells and whistles" were replaced by simple knobs with pictures. In order to heat a cup of soup, it was no longer necessary to follow a series of complicated instructions regarding how to control the intensity and duration of the electromagnetic radiation; one had only to touch the little-picture-of-a-cup-of-soup button. One and the same technology thus had a new and gendered identity in a different context. The two microwaves were equally functional, intended to speed the preparation of meals, but the one defined its users as technologically competent, the other as incompetent.

In fact, the microwave has this gender-stereotyping effect not so much because of its *materiality* but primarily thanks to its *meaning.* The original microwave was a symbol for technological sophistication because of its complexity, while the newer version symbolizes technological illiteracy by virtue of the low-tech places where it is sold and its simplified operation. When the microwave is considered in terms of the ways in which it is present as a working *material object*—a quick and easy food-heater-upper—yet another role in its use context becomes visible: it appears to be able to change human eating habits. The microwave facilitates a particular kind of meal, the frozen, ready-made kind that can be "prepared" in a short period of time and for a single person. It promotes such meals amongst its users, thus fostering a change in eating habits in which fewer are taken in company and more are eaten solo.

An Oprah Winfrey program provides a striking illustration. Its producers asked a number of families to share dinner together every day for a period of time—families whose members had previously eaten separately on their own time, who were now being asked to break the habit. Several families appeared to find the experiment difficult to complete, though a number of heart-warming scenes did occur. The father of one participant, for instance, thanked Oprah profusely, saying that without this experiment he would never have known that his son wanted to join the Navy. Evidently, in at least one household, the microwave had done much more than warm food.

Such aspects of technology can only become visible when technology is considered from the point of view of its concrete artifacts. These remain invisible when technology is only conceived as springing from a particular manner of thinking or from the functional organization of modern social life. The microwave then becomes only one of the many technologies that express that manner of thinking or that require that particular functionalism, whereas its actual role in our daily lives appears to comprise much more than this. The differentiated picture that empirical studies of technology provide of the role of technology in human existence and of the experiences of human beings thus demands a rethinking of how this role needs to be understood philosophically. The challenge posed by empirical studies of technology to the philosophy of technology is to understand technology not only in terms of its conditions of possibility but in terms of concrete artifacts, and yet to continue to pose philosophical, and not merely empirical, questions.

The Orphic Temptation

In this book I shall attempt to respond to this challenge by outlining another way of thinking, another direction for the philosophy of technology. In a certain sense, traditional philosophy of technology approached its subject matter from a *transcendental* direction. Transcendental philosophy, which achieved its zenith in the work of Immanuel Kant, takes as its point of departure the analysis of conditions of possibility. A transcendental-philosophical theory of knowledge, for instance, consists of the elucidation of the conditions of possibility for knowledge; it attempts to bring out into the open everything that must be supposed in order for knowledge to be possible. The presuppositions that are brought to light in this way are not empirically observable but "transcendental"; that is, they overstep or transcend (cannot be found in) empirical reality, but nevertheless must be presupposed in order to understand reality.

Classical philosophy of technology proceeded "in the style of" transcendental philosophy, in the words of Maarten Coolen (Coolen 1992, 108), though to be sure the conditions of possibility that it uncovered were not truly "transcendental" (in the sense of nonempirical) in nature, as we shall see. Classical philosophy of technology tried to understand technology from its conditions of possibility, from what must be presupposed in order for it to be possible. It thought "backward," so to speak; from the actual presence of concrete technological objects in our society to what made them possible.

This approach has produced many relevant insights and to a large extent has shaped the understanding of technology and its role in contemporary culture. But our picture of technology is distorted if technology is approached *exclusively* in terms of its conditions of possibility. For then we are speaking about technology's conditions of possibility as if we were speaking about concrete technologies themselves, and the transcendental perspective becomes absolutized into *transcendentalism.*[3] This is precisely what happens in classical philosophy of technology. Whenever technology is exclusively defined in terms of its presuppositions, aspects of the picture drop out that can only appear in a more empirically oriented approach, one that investigates the role played by specific technologies in specific contexts. Classical philosophy of technology prestructured its analysis in such a way that it could not *but* discover alienation. It failed to see that the diagnosis

3. The term "transcendentalism" is often used to characterize the views of classical philosophers of technology, Heidegger's in particular. To my knowledge, two authors have referred to Heidegger's thinking as "transcendental": Egbert Schuurman and Maarten Coolen. In his book *Technology and the Future: A Philosophical Challenge,* Schuurman uses the term "transcendentalists" to refer to philosophers who "investigate the transcendental directedness of all experience and thus also of technological actuality" and who "pay attention to the relative autonomy of technology and its tendency to absoluteness" (Schuurman 1980, 60). Heidegger, for him, is one such thinker. In his book *De machine voorbij* (*Beyond the Machine*), Coolen characterizes "the technological approach" to technology as one that analyzes technology as "a type of comportment (vis-à-vis nature) which is necessary for human beings

that technology *presupposes* a dominating manner of thinking or a functional orientation of social life does not necessarily imply that dealing with concrete technologies can only *produce* this domination and functionalism.

Classical philosophy of technology, therefore, met the same fate as befell Orpheus in Greek mythology, who attempted to lead his beloved Eurydice out of Hades but who was not permitted to look back in the process. At the last moment, on the threshold of the ordinary world and despite all warnings, he glanced around to catch a glimpse of her, only to see her disappear forever. The philosophy of technology needs to resist this "Orphic temptation" of looking backward. It must be confident that it will be able to get a full view of technology once it has left the realm of the transcendental and reenters the world of concrete materiality.

The charge mentioned above that classical philosophy made claims both too abstract and sweeping flows forth from its transcendentalist springs. Technology cannot be reduced without remainder to what underlies it. When Heidegger, for instance, conceives of technology as a dominating and controlling way of thinking and engaging with the world, and ultimately as a specific manner of world-disclosure or "being," he opens up an important perspective on technology. But this perspective is not sufficient to adequately analyze concrete technologies. To say that technologies *spring from* a certain manner of thinking and comporting oneself—their manufacture, after all, requires a certain manipulative intervention into reality—does not mean that such a manner of thinking and comporting is the only *allowable consequence* of using technologies.

When dealing with technologies, much more happens than object manipulation. A person who sends an email does not, after all, treat the addressee as raw material, nor does a passenger on a train so treat the landscape. The question of the kind of way that an addressee is present to the sender of an email, and the landscape to a railroad passenger, is precisely the kind of question addressed by the "new style" philosophy of technology. This new philosophy is not a backward but a forward thinking, starting from the technologies themselves and asking what role they play in our culture and daily lives, instead of reducing them to the

to be able to have technical instruments at their disposal." Coolen calls Heidegger "the great example" of such an approach, for he understands technology "as an essential ingredient of an all-embracing metaphysical comportment towards all beings that holds sway in our culture." This means Heidegger's position is not merely transcendental but transcendentalist, "for it sometimes seems as though Heidegger denies that human beings can ever escape their imprisonment within the comportment of modern technology" (Coolen 1992, 107–12). I agree with Coolen in including Heidegger among those who think in the style of transcendental philosophy, but to his reasons for calling Heidegger a transcendental*ist* I would like to add that it is not so much that Heidegger sees technology's conditions of possibility as independent, but rather that he confuses them with technology itself, as I show in Chapter 2.

conditions of their possibility and speaking about these conditions as if we were speaking about technology itself.

As the Belgian philosopher of technology Gilbert Hottois observed, the phenomenological tradition of the philosophy of technology to which Heidegger belonged is technophobic. It failed to notice the unique and radically new character of modern science and technology—which he characterizes as "technoscience"—and tried instead to shoehorn them into its existing conceptual frameworks. The phenomenological tradition, or at least the work in the philosophy of technology that takes its cue from Heidegger, conceives of technoscience as a particular kind of *interpretation* of reality, and fails to see its *operativity*, which makes it transcend the realm of interpretation (Hottois 1996a). The technosciences are more than interpretations of reality; they act, even encroach, upon reality. We fail to understand technology adequately if we only characterize it in terms of interpretation, for this reduces it to the domain of the symbolic, which is what it precisely transcends. This is not to say that approaches such as Heidegger's should be considered entirely obsolete. On the contrary, Heidegger and Jaspers have drawn attention to an important dimension of technology; namely, the relation between technology and the way in which human beings interpret and engage their world. That perspective, however, can be filled in much better by seeking a closer contact with technology itself, which is precisely what becomes possible via an analysis of technology in terms of its concrete artifacts. This would not reduce technology to something nontechnological, but instead would describe it in terms of its concrete presence and reality in human experience and practices. Such an approach—which looks at concrete technologies with an eye to the relations between human beings and world that they make possible and elucidates the structure of these relations—would substitute forward for backward thinking. It then would become possible for the philosophy of technology to raise philosophical questions and issues while at the same time taking seriously empirical investigations into technology.

TOWARD A PHILOSOPHY OF ARTIFACTS

In this book, I outline the elements of a "philosophy of technological artifacts," and bring it to bear upon issues of industrial design. In order to do so adequately, however, I first point out which problems in the philosophy of technology this "turn to artifacts" is meant to solve. The two chapters in Part I, Philosophy Beyond Things, therefore are devoted to a critical analysis of the way in which classical philosophy of technology has approached its subject.

The first classical conception of the influence that technology has had on human existence that I analyze is in the work of Karl Jaspers. His existential approach emphasizes the idea that technology creates large-scale "mass rule," fostering mass production and mass culture and promoting an entirely new way of existence in an entirely new material environment. Jaspers interprets this development as bringing about the alienation of human beings from the world and from themselves: technology entails a loss of "authenticity." Jaspers proposes that to overcome this alienation human beings need to realize that technology is ultimately only a neutral means for achieving goals that they themselves set. Were this to happen, coming to terms with technology would be seen as a *task* or *challenge* for human existence rather than as *undermining* it.

The second classical conception of technology that I shall analyze is that of Martin Heidegger, whose approach highlights the technological relationship of human beings to the world. Heidegger understands technology as a particular manner of approaching reality, a dominating and controlling one in which reality can only appear as raw material to be manipulated. In analyzing the views of Jaspers and Heidegger on technology in Chapters 1 and 2, I will be examining in particular the connection between their diagnoses of alienation and the ways in which they conceptualize technology. This will show not only how classical philosophy of technology was beholden to the Orphic or transcendental temptation of thinking backward, but also the cost it paid for so doing.

The work of Jaspers and Heidegger represents the two poles of the phenomenological tradition within the philosophy of technology: *existential phenomenology*, in which the central question is how human beings realize their existence and thus are present in their world, and *hermeneutical phenomenology*, which examines the ways in which reality is interpreted and thus is present for human beings. But not only have phenomenology's offspring in the philosophy of technology come under fire, so has phenomenology itself, by virtue of the romantic and essentialist manner in which its investigations are sometimes carried out. These investigations sometimes arouse suspicion that phenomenology strives for "authentic" contact with "reality itself," to contrast that with the alleged alienation produced by science and technology.

Yet a recent development in the phenomenological tradition, which I follow Don Ihde in calling "postphenomenology," offers a preeminent possibility of formulating a philosophy of technological artifacts of the sort that I think is necessary for a philosophical rethinking of artifacts and materiality. In Chapter 3 I shall set out the elements of this postphenomenological perspective and illustrate how it makes possible an alternative working out of the existential and hermeneutical approach to technology that liberates phenomenology from its problematic elaborations. I shall show how it can provide a framework for understanding the role of artifacts in the practices and experience of human beings; the ways in which human beings can be present to their world, and the ways in which the world can be present to them. This framework makes it possible to develop a new implementation of the existential and hermeneutical perspective on technology.

In Part II of this book, Philosophy From Things, I continue to elaborate this "postphenomenological philosophy of technology" via a critical discussion of the work of contemporary thinkers including Don Ihde, Bruno Latour, and Albert Borgmann. In contrast with the transcendentalism of classical philosophy of technology, I articulate an approach to technological culture that attempts to understand the concrete role of technological artifacts in human existence. The key concept of this approach is "mediation." This concept allows us to escape from two "common-sense" approaches to technology, which regularly frame the discussion, and which, as we shall see, can be found in the work of Jaspers and Heidegger as well. The first is the *instrumentalist* view that technology is a neutral means to achieve human goals be they good or evil; the second is the *substantivist* conception that technology is not neutral but a determining and controlling influence on society and culture. When technological artifacts are looked at in terms of mediation—how they mediate the relation between humans and their world, amongst human beings, and between humans and technology itself—technologies can no longer be pigeonholed simply as either neutral or determining.

On the one hand, the concept of mediation helps to show that technologies actively shape the character of human-world relations. Human contact with reality is always mediated, and technologies offer one possible form of mediation. On the other hand, it means that any particular mediation can only arise within specific contexts of use and interpretation. Technologies do not control processes of mediation all by themselves, for the forms of mediation are always context-dependent—otherwise we would be back at the technological determinist view. To return to the example of microwave ovens, these can only change the eating habits of human beings under two conditions: first, when frozen foods and other easily reheatable meals are available; and second, when human beings are prepared to adopt these as an alternative to preparing a meal from scratch with fresh ingredients. In the presence of these two conditions, microwave ovens do indeed shape the social context in which they function; without them, this context would be much different.

In the third and last part of this book, Philosophy For Things, I bring the "philosophy of artifacts" to bear upon industrial design. An approach to technology in terms of artifacts makes this connection not only possible but inevitable, and the philosophy of technology thus becomes directly relevant to technological practice. Industrial designers produce artifacts that embody a mediating role in the daily lives of human beings, and the philosophy of technological artifacts could help to explicitly anticipate the forms of mediation. In order to make room for this anticipation in industrial design, I shall use the postphenomenological perspective to develop a comprehensive framework in which to understand the role of aesthetics in design. The contemporary emphasis on symbols and lifestyles that I mentioned at the beginning of this introduction must be broadened to include the anticipation of the mediating roles of products—the ways they coshape the sensorial contact between humans and their world.

As a point of departure for considering what a "material aesthetics" that anticipates the mediating role of things can mean in practice, I examine the work of the Dutch Industrial Designers' Association, Eternally Yours. Eternally Yours follows an unorthodox approach within ecodesign. Instead of the usual emphasis on reducing pollution while maintaining beauty and economy, the company focuses on lengthening what it calls the product's "psychological lifetime." Most products are thrown away long before they are broken or obsolete, usually because of changing tastes and fashions. Eternally Yours attempts to combat this tendency of products to wind up prematurely in the landfill by designing products that invite people to become attached to them. Eternally Yours strives to achieve what the Italian designer Ezio Manzini calls "caring for objects." Products must be allowed the possibility to "grow old in a dignified way," and so to break out of our implicit cultural assumption that artifacts only have a limited lifetime and instrumental value. In order to stimulate this kind of attachment, as I show, one first has to analyze the relations between human beings and artifacts. The postphenomenological perspective can make an important contribution in precisely this area.

Edmund Husserl provided the watchword for phenomenology with his call, "To the things themselves!" In this book I heed his call literally. What holds for phenomenology holds equally for the philosophy of technology and for industrial design: To the things themselves!

PART I

Philosophy Beyond Things

1

Technology and the Self

INTRODUCTION

One of the most important issues addressed by classical philosophers of technology is the transformation of society into a mass culture, which is marked by the homogenization of the world and of the human beings who live in it. Technological developments since the Industrial Revolution, these thinkers have pointed out, have given rise to a system of mass production whose yield is a disturbingly uniform array of consumer products and whose requirements in the way of work and organization treat human beings primarily as interchangeable productive forces, insignificant cogs of mass society. These developments are making it ever more difficult for human beings to exist as unique individuals and interfere with their ability to authentically realize their own existence.

Jacques Ellul, for instance, has used the term "universalism" to refer to that feature of modern technology wherein it is applied worldwide, irrespective of place and culture. In the modern world Ellul detected the birth of a universe in which technology was the common language, in which society was being transformed into a single system that held out no place for human beings as subjects, and in which human beings were defined only in relation to that system (Tijmes 1992a, 55; 59). Günther Anders has emphasized the homogenization of the material environment created by the system of mass production. Given that each mass-produced artifact exists in the form of countless identical copies, any specific object human beings deal with is of less and less importance (Anders 1987, part 2, 37). And Walter Benjamin, to mention a final example, writes of the "adjustment of reality to the masses and of the masses to reality" (Benjamin 1968, 223).

Approaches such as these, which are concerned with the role technology plays in human existence, describe technology implicitly from an existential perspective. The aim of existential philosophy is to understand the nature of human existence. Though thinkers of the existential philosophical tradition have often dealt with the question of technology implicitly, they have only rarely dealt with it explicitly—which perhaps explains why relatively little attention has been paid to the existential tradition in many surveys of the philosophy of technology.[1] One existential philosopher who does raise the question of technology explicitly is the German thinker Karl Jaspers, who discusses technology extensively in several of his works. His all-too-neglected philosophy of technology, over half a century old, is the most systematic example of the classical existential approach to technology.

While I will show that better answers can be found than the ones Jaspers provided, the questions he introduced are no less urgent today. His views are vulnerable to criticism on several counts, yet they are worth close examination. By critically analyzing the ways in which Jaspers conceptualizes technology, I aim to expose the reasons behind the vulnerabilities in his elaboration of the existential perspective. This systematic investigation of the relation between Jaspers's approach to technology and the problems to which it leads will clear a space for a new articulation of the existential perspective, to be elaborated more fully in the second part of this book.

Jaspers developed his philosophy of technology in two phases. In his early conception of technology, which he put forth in 1931 in *Man in the Modern Age*, his central thought revolved around the transformation of human society into a mass, mechanized culture. Technology, he felt, posed a threat to what he called "the authentically human"; technology alienates human beings from the possibility of existing as unique individuals and turns them into anonymous accessories of mass culture. Jaspers's later conception of technology, in which he looked more closely at the nature of the threat posed by technology and into how it might be overcome, took a more ambivalent stance. He concluded that tech-

1. See, for instance, Achterhuis (2001); Mitcham (1994); and Rapp (1994).

nology is ultimately neutral in itself, no more than a means for human goals, for it is incapable of generating its own goals. This neutrality makes human beings responsible for what they make of technology. While technology may indeed be a threat to human existence, the task of recovering their mastery of it at the same time offers human beings the chance to regain their humanity.

In order to analyze Jaspers's philosophy of technology I shall first describe his understanding of the relation between technology and modern life, then critique two of its elements: the new and homogenized material environment that Jaspers sees as the fruit of mass production, and the mass culture that forms the new and homogenized social environment. This will put me in a position to broach the second phase of his philosophy of technology, and evaluate his position in the classical existential approach to technology.

TECHNOLOGY AND MASS RULE

Jaspers's early view of technology can be summarized succinctly: technology suffocates human existence. The technological developments of the past two centuries, he thought, have radically changed human life. Up through the eighteenth century the role of technology in society was relatively limited and clearly defined. It was mechanical in character and, though originally driven by muscle power, it was eventually augmented with animal, wind, fire, wind, and water power. At the end of the eighteenth century, during the Industrial Revolution, this limitation was shattered with the birth of modern technology in the industrialization of production processes, which according to Jaspers had a completely different character than traditional technology.

From that time on, technology has come to play a role in virtually every aspect of human life. Jaspers's initial estimation of this transformation was extremely negative. In *Man in the Modern Age* he spoke of the "demonism of technology," describing technology as a demonic, independent power that had been summoned into existence by human beings but that now has turned against them. The demonism of technology consists in its transformation of human society into what Jaspers called "mass rule."

The Apparatus

Mass culture, according to Jaspers, is a byproduct of the interaction between technological development and population growth, which has been exponential in recent centuries. Jaspers marveled at the fact that there were 1.8 billion people alive in 1931, while there had only been 850 million people living in 1800; today, the world population has grown to over 6 billion. This population explosion would have been impossible without such technological developments as more efficient modes of production and work organization, clinical medicine, and quicker and more extensive transportation networks. One consequence, however, is that the vast number of human beings whose existence has been made possible by technology are now utterly dependent on it: "The broad masses of the population could not exist today but for the titanic interlocking wheel-work of which each worker is one of the cogs" (Jaspers 1951, 39).

But Jaspers sees this dependence on technology as only one of the prices that humanity must pay for its vast numbers. The population growth that technology has made possible also depends on a quite specific social and cultural formation. Paradoxically, technology gives us more work to do in lightening our work load—for it creates more needs than it fulfills, thanks to the population expansion it produced and to the creation of new commodities that, in turn, create new needs. But the new work that technology stimulates is associated with the *mechanization* of labor. Machines are more frequently employed in manufacturing in order to produce commodities with greater rapidity and efficiency—meaning that work is less and less involved with the personal manufacture of artifacts and more and more with the maintenance and operation of machines that make parts for artifacts. The growing work capabilities require the development of a smoothly operating organization, leading to the creation of an extensive *bureaucracy.*

Supplying the needs of the sharply expanding world population therefore requires a complicated interplay between mechanization and social organization. Everything must be planned and coordinated with everything else. The tightly organized society that results, according to Jaspers, itself has the character of a machine. He therefore describes technological society as "the Apparatus." The ongoing development of technology has given rise to a gigantic apparatus, which organizes and outfits the whole of social life, consisting of workers, machines, and bureaucracy, upon which all of these are dependent. The apparatus increasingly determines how human beings carry out their daily lives. This leads to what Jaspers calls "mass rule" or "mass order," and has two different but related effects. First, the apparatus creates a system of mass production that fosters a homogenization of the material environment in which human beings live. Second, it approaches human beings not as unique individuals but as fulfillers of functions who are in principle interchangeable. I shall speak about each of these two elements of Jaspers's theory of mass order in turn.

Mass Production

According to Jaspers, technology has enormous implications for humans' material environment. Thanks to mass production, for instance, it poses a threat to the "bond between human beings and the world." Traditional technology was on too small a scale to threaten this bond, and human beings played a large role in the production process. Ever since the rise of modern technology, however, humans have produced less and less themselves. The apparatus, through mass production, delivers an abundance of commodities that can be consumed instantly to satisfy immediate desires without human beings having played a significant role in bringing them about.[2]

> The upshot of technical advances as far as everyday life is concerned has been that there is a trustworthy supply of necessaries, but in a way which makes us take less pleasure in them, because they come to us as a matter of course instead of with the relish given by a sense of positive fulfilment. Being more materials obtainable at a moment's notice in exchange for money, they lack the aroma of that which is produced by personal effort. Articles of consumption are supplied in mass and are used up, their refuse being thrown away; they are readily interchangeable, one specimen being as good as another. In manufactured articles turned out in large quantities, no attempt is made to achieve a unique and precious quality, to produce something whose individuality makes it transcend fashion, something that will be carefully cherished. An article which thus satisfies ordinary needs arouses no peculiar sense of affection. (47–48)

The "bond" between humans and artifacts is hampered not only by a lack of involvement in their production, but also because the artifacts evoke a minimum of affection when used. Technologically produced artifacts, according to Jaspers, tend to evolve into "ideal types" with a standardized function and a mass-producible form.

2. Moreover, Jaspers claims in his *Philosophy* that there can never be a *fully* produced world. There is always a tension between plans and results, thanks to the role that local conditions, human peculiarities, and accidental circumstances play in the technological outfitting of the world. Because of these, there will never arise a completely closed world of consumers and consumables (Jaspers 1969–71, 1:77). Jaspers's diagnosis, therefore, should be seen as an account of a tendency rather than as a description of a factual state of affairs.

> Among articles of consumption we distinguish the well-adapted and substantially perfected kinds, the definitive forms whose manufacture has become thoroughly normalized. . . . When perfectionment has gone as far as this, fondness for a particular specimen has become unmeaning. The general form is what matters to us, and, however artificial that may be, such things have a functional suitability which almost makes them seem like natural products rather than the creatures of man's activity. (48)

The removal of any bond between human beings and their world has a tremendous impact on the way in which humans can give form to their existence. For human beings, according to Jaspers, need such an attachment with the world in order to realize their individuality. Only then can *the* world become *their* world—an environment that allows not only the anonymous functioning of parts but also personal engagement and commitment; an environment in which human beings not only satisfy their needs but also realize themselves as authentic individuals. They will be unable to do so, for Jaspers, in a world consisting of merely useful objects. In that case, their world holds them captive: "All live alike, in the same worldless satisfaction of needs by identically replaceable things and materials; all are completely dependent upon each other for their concrete means of existence, yet without necessarily being in personal touch. The only freedom left to men by the calculable course of this endless productive machinery would be the freedom to watch" (Jaspers 1969–71, 1:112–13).

The Masses

The transformation of human existence into mass rule, according to Jaspers, was considerably strengthened by the way in which the apparatus facilitated "the dissolution of the individual into his functionality." In the fully organized and outfitted social life created by the apparatus, it does not matter who you are but rather what your function is; human beings are stripped of their personal uniqueness not only by their material environment but also by their social environment. Within the apparatus everyone is, in principle, interchangeable. Differences of age or character become irrelevant: "The individual is no more than one instance among millions; why then should he think his doings of any importance?" (Jaspers 1951, 50). This functional conception of human beings also strips them of the possibility of realizing their authentic personal existence in the workplace. Within the apparatus, "the day's work grew sufficient to itself and ceased to be built up into a constituent of the worker's life" (45). Work comes to be distinguished from "free time," which acquires the character of "pastime," for people use it to rest from their work. "This positive gratification of the mind without personal participation or effort promotes efficiency for the daily round, fatigue and recreation being regularized" (50).

The apparatus therefore reduces human beings and their material environment alike to their functioning. "The [modern attitude of mind] does not want phrase-making, but knowledge; not ponderings about meaning, but dextrous action; not feelings, but objectivity; not a study of mysterious influences, but a clear ascertainment of facts. . . . Control and organization are supreme" (49). The result is a leveled-off culture: "Essential Humanity is reduced to the general" (49).

The shape that human existence takes in the form given it by the apparatus Jaspers calls "mass life." In mass life, human beings are no longer capable of authentically "being themselves." Technology threatens humanity at its core by its tendency to reduce humans to their functions and to stifle their interactions with their environment. Their social environment allows them a place within the apparatus that could also be taken by others. Their material environment acquires—just as they themselves do—a more and more functional character, with which a personal bond is less and less possible. The space that humans have to realize their individuality becomes ever smaller. Personal uniqueness gives way more and more to impersonal interchangeability. In *The Origin and Goal of History*, Jaspers provides a bitter and rather elitist sketch of "the mass":[3]

> The mass . . . is not subdivided, is unconscious of itself, uniform and quantitative, devoid of specific character and cultural heritage, without foundations and empty. It is the object of propaganda, destitute of responsibility, and lives at the lowest level of consciousness. . . . Masses arise where men come to be without an authentic world, without provenance or roots, disposable and exchangeable. In consequence of technology this state of affairs is growing more and more widespread: the narrowed horizon, life that does not look ahead and is devoid of effective recollection, the compulsion of meaningless labor, amusement in the dissipation of leisure, excitation of the nerves masquerading as life. (Jaspers 1953, 128)

3. Elisabeth Hybašek, in her analysis of the relation between Jaspers's approach to mass culture and his vision of industrial society, points out that his description of "the mass" draws heavily from Gustave le Bon's *Psychologie des foules* of 1895, which surely by 1949, when Jaspers wrote this work, was considered obsolete and elitist (Hybašek 1984, 245–50).

Mass life for Jaspers thus not only is the occasion for the onset of the apparatus in a quantitative sense—the rising numbers of human beings on the planet need technology to exist—but is also brought forth by it in a qualitative sense: the apparatus only allows humans to be present as anonymous parts of "the mass." Herein lies the demonic character of technology: without it being the explicit intention, the development of technology resulted in an explosive population growth that demanded that society be organized as an apparatus—with the downside that the apparatus brings with it a specific kind of social existence that Jaspers called "mass culture."

The threat that technology makes to the possibility of human beings to live as authentic individuals leads Jaspers to say that the task of humanity is to recover the possibility of selfhood—of being oneself. This does not entail a complete repudiation of technology. Technology forms the world in which humans live, from which they cannot withdraw. Therefore, Jaspers writes, "because selfhood exists only in unity with the being of the time, it is still resolutely determined to live only in this same time even though it should find itself in conflict therewith" (Jaspers 1951, 196). He emphasizes the importance of achieving a "new nearness to the world," and ultimately of engaging oneself unconditionally with it. Only from such an engagement can human life find its destiny rather than be "the life that proceeds in a succession of passing moments until it comes to its term" (182). Existential engagement transforms reality into the meaningful place of one's existence: this engagement provides the room in which human beings can exist "as themselves." Humans have to "immerse" themselves in the world (183–84), rebelling against our contemporary "remoteness" from it and devoting themselves to the other people and things in it.

This new presence of things can only arise when technology is used in a resigned or reduced manner, and when we restrict ourselves to what is absolutely necessary. Such an attitude allows space for another kind of nearness to things, a person's openness to something other than the mechanistic, "so that he can reflect about, so that he can allow to ripen, so that he can enter into genuine contact with, the things which are his own" (180). In *Man in the Modern Age,* Jaspers cautiously proposes that technology could play a role in this new engagement with the world. It can, that is, allow us new kinds of access to the world, for instance by enlarging our perceptual abilities or by giving us new modes of transportation that enlarge rather than shrink our world (e.g., 179–80). But these are only passing remarks. In Jaspers's early work, the view that predominates is that technology alienates human beings from themselves and from the world.

HUMAN BEINGS AND MASS PRODUCTION

Before broaching Jaspers's later philosophy of technology I shall, in the next two sections, elaborate further his account of the relation between technology and mass rule, as well as criticize his approach on a number of points. His analysis of the implications of the formation of a system of mass production for human existence is full of intriguing observations, but it ultimately boils down to a variant of the alienation thesis—that technology estranges human beings from reality and from themselves—which, as I have mentioned in the introduction, has been shown to be untenable by the philosophical developments of the past few decades. Based on an analysis of Jaspers's argument, I show that his gloomy diagnosis of the technological culture is a direct outgrowth of the particular way in which he elaborates the existential perspective on technology. His analysis is a clear example of the transcendentalist manner of thinking elaborated in the introduction, which reduces technology to its conditions of possibility and then speaks of these conditions of possibility as if it were speaking about technology itself. I also point out the extent to which Jaspers's analysis was shaped by the time in which he wrote. His position becomes more comprehensible when it is historically contextualized, for his statements about "Technology" are artifacts of a specific period in the development of technology.

In this section and the following one I shall elaborate the issue of the relation between technology and mass rule in two directions. I shall first describe and criticize the implications of technology for the *material environment* of human beings; then I shall describe and criticize his analysis of the consequences of technology for *human existence:* how it shapes society via "the Apparatus" and gives rise to mass culture.

The Relation Between Human Beings and Technological Products

One way in which technology threatens human existence, for Jaspers, is that it stifles the bond between humans and their world. In his view, mass production curtails the attachment between humans and the world around them in two ways: human beings no longer have a hand in producing the artifacts themselves, and the artifacts they produce are standardized and merely functional. Mass-produced artifacts are not valuable as individual objects; only their function matters, and this can also be fulfilled by other, identical copies.

On the one hand, these observations by Jaspers are enlightening and surprisingly contemporary, but on the other hand they are also highly problematic.

Jaspers was remarkably prescient in his concern with respect to the lack of a bond between human beings and industrial products. Only recently have similar concerns been voiced, though coming from a completely different angle: ecologically friendly industrial design. The Dutch Industrial Designers' Association, Eternally Yours, mentioned in the introduction as concerned to develop ecologically sound products, regards the strengthening of this bond as its most important task in reducing the environmental burden. When human beings have such a bond with products, Eternally Yours finds, they are less inclined to throw products away prematurely, i.e., before they are actually obsolete. This is much more effective from the point of view of environmental protection than boosting energy efficiency and reducing waste emissions in the production and use of artifacts, since many products appear to be discarded long before they are worn out.

Jaspers's diagnosis of the implications of technology for the relation between human beings and their material environment, however, contains a number of problems, which arise from his assumption that mass-produced artifacts are *only* functionally present for human beings. This assumption invites two criticisms.

The first is that his remarks are not really directed to products as such but rather to functionalism, a specific phase in the history of industrial design. Functionalism was the watchword of a style that called itself "modernism" and that flourished in the 1930s, precisely the time in which Jaspers wrote *Man in the Modern World.* A short recapitulation of the history of industrial design and its impact on material culture makes it possible to contextualize Jaspers's analysis and to get a better view of how our material culture has fared since it was taken in serial production.[4]

When manufacturing processes were mechanized, the first reaction of practitioners of the "applied arts" was aversion. In the second half of the nineteenth century their protest gave rise to the "Arts and Crafts Movement," which explicitly opposed the encroaching mechanization and which advocated a return to traditional craft production methods. This movement, with the utopian William Morris as a figurehead and "a thing of beauty is a joy forever" as a slogan, celebrated the making of artifacts that presented an alternative to standardization and to mechanical production methods, in which workers were alienated from their products. But the resistance to the machine quickly faltered. After a short transition period represented by art nouveau (*Jugendstil*)—a movement from the beginning of the twentieth century that was less fiercely opposed to mass production but that did

4. In what follows I have relied upon Heskett (1980); McDermott (1992); Ramakers (1986); and Sparke (1986).

extensively adorn its objects with the images of plants, insects, and female nudes as a way of referring to nature—mechanization gradually won out over the romantics. Partly under the influence of the socialist ideal that each individual in society was owed both a worthy job and a worthy standard of living, the machine gradually came to be embraced rather than repudiated. The mechanization of the manufacturing process, after all, made it possible to produce great quantities of affordable and substantial commodities, and at the same time seemed to free human beings from the need to labor at menial tasks.

Modernism, which flourished in the 1930s, was the industrial design movement that grew out of these ideas. So enthusiastically did modernism embrace the machine that it even became the model for the products it was used to manufacture: just like machines, products were supposed to be *functional.* "Form follows function" quickly became the slogan and the most important characteristic of modernism. Le Corbusier, for example, called chairs "sitting machines" and houses "machines for living." Furthermore, not only the functionality but also the aesthetics of the machine became a model for designers; whence the simple geometrical forms and the frequent and conspicuous use of tubular steel in furniture of this period. A no-nonsense design tradition arose wherein the characteristics of products were modeled after the machines with which they were made. In this first main period of the development of industrial design, an entirely new generation of products came into being whose design was composed mainly with an eye to their functionality. And this is precisely the period in which Jaspers formulated his critique of the purely functional character of the contemporary material culture.

But design history did not stop with functionalism. Gradually, consumers, not products, became the center of the designers' attention. Functionalism became watered down into fashion: it was translated into decoration and "styling." The most recognizable example of this development is "streamlining," in which products were given the outward shape of modern machines, especially airplanes. Art deco, too, was a manifestation of this "weak modernism." It translated the modernist belief in technology into decorations that invoked or referred to it, such as geometrical shapes or electrical sparks. This stress on the appearance of products was completely at odds with the functionalist conviction that objects should be designed only in view of their function; if modernism loathed anything, it was decoration and stylization.

Modernism remained influential until after World War II. In the 1950s a new movement arose, the "contemporary" style, which reflected the optimism fed by the postwar economic boom. New materials, appliances, and production techniques were developed and used with great fanfare. In many respects, the contemporary movement was a continuation of the quest for functional and indestructible products. Nevertheless, it amounted to a step back from modernism, for it was explicitly developed as a style, with its famous chemically developed colors such as hot pink and its light furniture with tubular metal legs. The increasing emphasis on style revealed that modernism was slowly crumbling.

The definitive collapse of modernism occurred in the 1960s with the rise of "pop" design. Pop was about as far removed from the modernist emphasis on functionality as one could get. It put the consumer front and center instead of the functionality and durability of the product. In its blatant appeal to the surging youth culture, it celebrated the trendy over the durable and the clever over the functional. A new aesthetic arose that designer Reyner Banham has named the "Throw-Away Aesthetic" (Banham 1981, 90–93). Pop paved the way for the decisive break with modernism, which took place in the 1970s with the coming of postmodernism.

Like pop, postmodern design rejected the modernist emphasis on functionality and durability. It saw products no longer merely as machines that fulfilled functions, but also, and above all, as embodying *meanings*. Products are used not only as instruments, but also to refer to lifestyles and social status. One purchases a couch not only in order to sit more comfortably, but also because it suits who one is or how one wants to be seen by others. And because of the enormous number of subcultures and lifestyles in contemporary culture, it is no surprise that this has spawned an immense diversity of styles; postmodernism rejects the one-dimensionality of modernism, and is less a new style than an explosion of styles. "Form follows function" was replaced by "form follows fun." A postmodern product exists by virtue not of its being a functional object, but rather of its being a bearer of meanings.

The strength of Jaspers's account of mass products is that he approached them in terms of the relations human beings have to them. His observation, for instance, that we have a quite specific kind of involvement with artifacts that are primarily functional is insightful. But his conclusion that the emphasis on the function of a product is intrinsically bound up with "the" mass-production practices of "modern technology" is much too hasty—for, as we have seen, in contemporary postmodern design mass production has led to an enormous diversity of material products, in which emphasis is placed not on *function* but on *style.* Far from settling into "ideal types," as Jaspers claimed, products are continually being reinvented in myriad ways. And they are not bought for their functionality alone but also for their beauty, or because they suit a particular lifestyle. This is not to say that Jaspers's concern about the dwindling bond between human beings and their material environment is entirely obsolete—the work of the Dutch association Eternally Yours testifies that it is not—but only that his analysis of the relation between human beings and their material environment that technology fosters must be seen in the perspective of the time in which he wrote.

Copies Without Originals

The second criticism I want to make of Jaspers's evaluation of the material culture of technology is that the reasoning on which he bases it does not justify his conclusions. The classical philosophy of technology, as already mentioned, frequently uses a transcendentalist style of reasoning that attempts to understand phenomena in terms of their conditions of possibility. One of my goals in analyzing thinkers from this tradition is to expose the connection between this line of reasoning and their typical thesis that technology leads to alienation. Jaspers's analysis of the material culture of technology offers an initial opportunity to do this. But in order to make this clear, I will first provide a short account of a pair of other perspectives on mass production that are closely related with Jaspers's; namely, those of the German philosophers Günther Anders and Walter Benjamin.

In 1978, Anders published an essay titled *Die Antiquiertheit des Materialismus* (*The Obsolescence of Materialism*) defending the claim that it is a mistake to describe the contemporary world as materialistic (Anders 1987, 2:37), for matter has become utterly irrelevant. According to him, we live in a second Platonic age, for at present "the individual object has a lower degree of being than its 'idea'; that is, its blueprint." What matters, he says, is not the object but only the ideas it embodies. "Of what importance is light bulb number 7846539-30 in comparison with its non-physical model?" Anders asks rhetorically. The answer: "no more than a reflection of an idea, a '*mè on,*' a non-being." Today, thanks to patenting, ideas can even be owned—something that did not occur to Plato even in his wildest dreams.

Anders had announced the death of the thing much earlier, in his 1958 *Die Antiquiertheit der Produkte* (*The Obsolescence of Products*). There he defended the claim, with a nod to Heidegger's concept of *Sein zum Tode,* or "being-toward-death," that our artifacts were "born to die" (Anders 1987, 2:38–57).[5] When artifacts are thrown into the world they are not in a viable state; we live in a throw-away world. Advertising constantly bombards us with inducements to

5. See also Dijk (2000).

buy products we already possess. Every day we have to face an unending "trash call" in which we are asked to exchange the artifacts we own for new ones. This leads to an "evaporation" of products. Products, after all, are supposed to posses "duration": a time to exist. But because they continually have to be replaced by other commodities, they are denied this duration. While for Plato things were inferior because they are perishable (in contrast to the eternal ideas), for contemporary industry objects are inferior because they do not perish rapidly enough. Anders names this relation between human beings and things the "negative ontology" of the industrial age.

In the background of Anders's critique is the idea that the new material culture involves alienation and a loss of authenticity. The "authentic" object of yore is something entirely different from the manufactured object of today that is but one among often thousands of material instantiations of the same. Whenever there are a large number of identical items, no real significance can be attached to any particular one. The thought that mass production leads to a loss of authenticity is drawn still more explicitly in Walter Benjamin's famous essay, "The Work of Art in the Age of Mechanical Reproduction" (1935), though his outlook was developed primarily in connection with artworks rather than everyday objects.

In "The Work of Art in the Age of Mechanical Reproduction" Benjamin inquires about what happens to the relation between human beings and works of art when these are mass produced—as, say, becomes possible through photography. His conclusion is that the mechanical reproduction of an object ruptures the "presence in time and space, [the] unique existence at the place where it happens to be" of the original. This "unique existence" safeguards the "authenticity" of the artwork, and this authenticity is precisely what is nonreproducible about it (Benjamin 1968, 220). Its authenticity confers upon the artwork an "aura" that is lost in reproduction: "By making many reproductions it substitutes a plurality of copies for a unique existence" (221). Benjamin lamented the steady encroachment of mechanical reproduction: "The adjustment of reality to the masses and of the masses to reality is a process of unlimited scope" (223).

It is not difficult to extend the argument that Benjamin makes here in connection with art works to everyday objects, especially given that, as is evident in the quotations above and elsewhere, he himself often speaks more generally in terms of mechanically produced "things." Everyday objects are mass produced as well, even more so than artworks: the average person who seeks a table today does not go to a craftsman to have one made by hand, but shops instead from an IKEA catalog. If there is any originality and uniqueness in the objects that surround us, it lies in the selection we have made from catalogs and stores rather than in the objects themselves, of which there are thousands of other identical examples. And the same compensation mechanism that Benjamin perceives in connection with artworks exists for artifacts as well. Just as media-driven "personality cults" supplement the lack of aura that the screen images of actors have in comparison to the stage actors who give live, "in person" performances in "genuine" theater, so advertising provides mass products with identity. The time is long past when the atmosphere of the workshop clung to artifacts, providing them with an aura of "authenticity" and "uniqueness."

There is, however, an important difference between mass products and the mechanically reproduced work of art that Benjamin writes about: the mass-produced object lacks an original. Mass-produced objects are not copies of an original but, to use Anders's words, copies of an idea. Ideas cannot possess auras, for they can never be experienced in the "here and now" and be "authentically" present. Mass products are not reproductions, and thus not reductions of an original. There is no original example of the chairs we sit on, the tables we eat at, or the pens we write with—or they are each equally "original." Any approach that depends upon the distinction between "original" and "copy," or "authentic" versus "alienated," therefore, is unsuitable for understanding the impact of mass production on contemporary material culture.

Anders and Benjamin are insightful in pointing out that mass production is a new phenomenon that fosters a new kind of material culture, but they fail to deliver a full analysis of the role that this new generation of artifacts plays in our daily lives. Again, in line with the transcendentalist tradition, they investigate the *origin* of mass products and on that basis draw conclusions about the places that these artifacts occupy in daily life. Anders's analysis of the mass-produced light bulb, for instance, approaches it only in terms of its role in the system of mass production by which it was produced, and not in terms of the role that it concretely plays in human life. From the fact that the manufacturing process produces a vast number of identical lamps Anders draws the conclusion that it is nonsense to attach any value to a separate, individual lamp. The reasoning is curious, for in order to understand what a mass-produced artifact means in human life one needs to analyze, not backward to how it *originated* or what its *conditions of possibility* were, but forward to what it actually *does.* The artifact itself must be looked at, rather than reduced to its origin.

This "backward thinking," which in the introduction I called "transcendentalism," also forms the backbone of Jaspers's critique of material culture. When Jaspers tries to understand the relation between human beings and the material surroundings shaped by technology, he does so not via an analysis of that relation or even of technology itself, but in terms of what has *made that relation possible* in its contemporary form: mass production. Jaspers reduces the material environment of human beings to its conditions of possibility, and then puts these conditions of possibility in the place of the material environment itself. To be sure, these conditions of possibility are not transcendental in the sense of being nonempirical, but I still think it is proper to designate them with that term for the reasons outlined in the introduction. The term suggests that concrete reality is transcended—and Jaspers indeed abstracts from concrete technology, albeit by reducing it to empirical phenomena like mass production or functionalism—and at the same time that this abstraction is then absolutized.

The problem with a transcendental line of reasoning is that, despite all of the enlightenment that it can provide, it is condemned to lose sight of those aspects of phenomena that cannot be reduced to conditions of possibility. That problem is clearly visible in Jaspers's analysis of the implications of mass production for the relation between human beings and artifacts. When artifacts are made possible by mass production and a functional design approach, that does not entail that the concrete role they go on to play in human life is merely anonymous and functional in nature. An individual can indeed be attached to a mass-produced and functional chair—for instance, if it is associated with particular memories, if it is thought to be pretty despite the existence of other identical examples, or if that individual has bonded to the chair over many years of use.[6] The relation between human beings and artifacts comprises more than what is visible when one only pays attention to the *origin* of the artifacts. An artifact can play more roles in human life than functional ones.

Jaspers's diagnosis is thus compromised by a category mistake. When the *conditions of possibility* of our material environment are clarified, it is still not by definition possible to say anything about our relation with the environment itself. Technology has drastically changed the human environment, but there remains much more to say than a transcendentalist perspective allows, more to say than that this leads to a functional and anonymous whole. To understand the role of technology in human existence, one must think not only backward to its conditions of possibility, but also forward to what technological artifacts themselves make possible, and what this means for human existence. Only when technology is thus investigated could conclusions of the sort Jaspers reaches be justified—

6. For a closer analysis of the attachment between individuals and artifacts, see Chapter 7.

yet, as I shall show in the second part of this book, this kind of investigation of technology leads to a quite different conclusion.

MASS EXISTENCE

The creation of a system of mass production, according to Jaspers, has had consequences not only for the material environment of human beings, but also for the structure of the society in which their lives take shape. Society, for him, becomes transformed into an apparatus that only allows people to be present in terms of their function rather than as unique, noninterchangeable human beings. The "mass existence" that results threatens what is most essential to human beings: their ability to "exist as themselves." In order to give a clearer picture of what Jaspers means by this, I shall provide a short introduction to his existential philosophy before critiquing his claim that technology transforms society into an apparatus in which only mass existence is possible, and, again, relate his gloomy vision on technology to the transcendental elements in his approach.

Existential Philosophy

Existential philosophy first appeared midway through the nineteenth century in the unorthodox work of the Danish thinker Søren Kierkegaard. Kierkegaard opposed the rigid system of the philosophy of his time, especially that of Hegel. Such a system, he felt, did not allow sufficient room for understanding what made human beings human. For human existence, Kierkegaard thought, "truth is subjectivity" rather than objectivity.[7] For Kierkegaard, and for the existential thinkers who came after him, human beings have a specific manner of existing; in their existing they are aware of themselves as existing. Human beings are not simply there, but *know* that they are, which makes their "being-there" something they have to realize actively. Their existence is not something simply given, but something that they have to shape themselves.

Existential philosophers have generally focused on clarifying the character of this specifically human way of existing. In Jaspers's work, existential philosophy crystallizes into a systematic whole. Jaspers delineates the specific character of the

7. The subtitle of section 2, chapter 2 of Kierkegaard's *Concluding Unscientific Postscript* is "Truth Is Subjectivity."

human way of being by contrasting it with two other, different modes of being: being as "Dasein," or "being-there" (Jaspers uses the word "Dasein" in a much different sense than Heidegger does, to refer to the thingly character of entities), and being as "transcendence." Every entity that can be an object of our thinking has the mode of being of Dasein, of something that "is *there.*" Jaspers calls the whole of what "is there" "the world," and the mode of being of what is there he also calls the "thingly mode of being."

Human beings, by contrast, "are" in a completely different manner. They are also thingly and thus have Dasein, and are part of the world, but there is much more to say about their mode of being. They also have an aspect that is utterly alien to other objects in the world, for they have the possibility of having a relation to themselves. Human beings do not coincide with themselves, but are *conscious of* the fact that they exist. And the fact that they have a relation to their own existence implies that they themselves give form to it, that they exist *themselves.* For Jaspers, therefore, existence is not an accomplishment but an ever-present task. It has to be given shape continually. Within the existential tradition, this human state of being is indicated as *freedom.* Their freedom gives humans the possibility not only of shaping their own existence, but at the same time of doing, or not doing, this *as themselves:* freedom offers the possibility of authenticity. When human beings act as themselves rather than simply react to stimuli they are not interchangeable instances of a certain type of being, but unique and authentic individuals.

Following Kierkegaard, Jaspers calls this specifically human mode of being, which is characterized by the possibility of freedom and authenticity, *existence,* or "the possibility of being [or becoming] oneself." "To be oneself" indicates both that "being" is something human beings have to *do* themselves (to *be or become* oneself) and that they can do this *as themselves* (to be or become *oneself*). By "being oneself," therefore, Jaspers does not mean something like "identity." This would misconstrue existence as a kind of encyclopedia of someone's character, inviting us to think of it as an object. Existential philosophy does not consist in an analysis of *who* someone is, but rather in the clarification of the human *mode* of being.

The twenty-first-century reader will probably find the emphasis that existential philosophers place on freedom and authenticity less convincing than did Kierkegaard and Jaspers. These concepts seem to ignore the situatedness and contextuality of human existence. Who human beings are, how free they are, which choices they can make, and even if it is possible at all to approach situations in terms of choices—all these questions depend on the contexts in which they find themselves. In the words of the Dutch philosopher Annemarie Mol,

"The ability to make choices is not a property of 'the human being.' Connections lie otherwise: situations of choice are always organized for specific people, on specific moments, and require a lot of effort" (Mol 1997, 8).

The practice of prenatal diagnostics gives a clear illustration of the fact that situations of choice are always *constructed* rather than being an intrinsic property of human life. Popkema and colleagues, for instance, have examined how the testing of embryos for Down's Syndrome creates a situation of choice that would not exist without the test: the choice whether or not to allow a child with that kind of condition to be born (Popkema et al. 1997, 114–15). Examples like this show that the autonomy of the choosing subject, which seems to be presupposed in Jaspers's existential philosophy, must be relativized. Mol prefers to speak of a "relational subject" instead of an "autonomous subject," which always stands in relations by virtue of which it becomes what it is (Mol 1997, 9). To be sure, Jaspers, too, relativized the autonomy of existence somewhat by discussing its historicity: existence has no fixed essence but is always becoming, and the process of becoming plays itself out in interchange with the circumstances in which human beings find themselves.

Yet such a relativizing of the subject's autonomy does not remove the fact that in these "constructed" situations of choice human beings can have the experience of acting "as themselves" or "not as themselves." Anyone who has a prenatal diagnostic test done knows that the possibility of a situation involving choice is created, in the event that the test shows that the fetus has a serious disease. The situation can be avoided by declining the test—but this, too, is a choice. Whatever the decision, the individuals involved will have to make it "as themselves." However much someone may discuss the decision with friends and family, ultimately one can never be content with having the decision imposed from the outside, with not making it oneself. That the subjects and objects of choice, and the situations calling for choice, are created rather than being like this "in themselves," does not alter the existential character of this configuration. In situations calling for a choice it is never a question of complete autonomy or contextless freedom, but neither of purely heteronomous relations and determinations. The fact that the existential dimensions of human existence are constituted rather than "natural" does not make these dimensions less existential.

The human possibility of "existing as oneself" becomes clearest, according to Jaspers, in situations that impose upon a person the realization that only he or she can give form to his or her own life. Jaspers calls these situations "boundary situations," such as experiences of guilt, suffering, and mortality. Such experiences make clear in forceful and unavoidable ways that human beings are what one might call "existers," individuals thrown back on themselves. Nobody can feel guilt or suffer for me; nobody can die for me. This "self" of the "exister" is the self that existential philosophy is concerned with, not the self in the sense of an autonomous subject that can realize itself at will in absolute freedom.

Alongside existence and being-there, Jaspers recognizes a third mode of being. When human beings explicitly experience their existential freedom, they immediately realize that it is not their own doing but a "gift" from a source they cannot grasp. They always already find themselves in their freedom; their freedom "happens" to them. This means that existence is not only a relating to oneself, but that this self-relation at the same time is a relation to something "beyond" it from whence it was "given." This "beyond," for Jaspers, possesses the third mode of being he discerns: "*transcendence.*" Transcendence is "being in itself." This designation appears paradoxical, for "being in itself" is always "being" independent of any particular appearance for a subject, and is thus in principle unknowable. It is impossible to know something "in itself"; as soon as we do, it becomes "for us." But just because being in itself is not *knowable* does not mean that human beings cannot *have a relation* to it: humans can establish *that* it withdraws from their understanding. "Being in itself" can be experienced as a boundary, as the ungraspable. For this reason Jaspers calls being "transcendence": it transcends our grasp, it is beyond understanding and manipulation.

One cannot say of transcendence that it "is there." It "exists" otherwise than do objects, it does not have the mode of being of Dasein. Therefore Jaspers says expressly that by "transcendence" he is not referring to a deity: actually, the belief in a deity is blind to transcendence, for it conceives transcendence as a thingly way of being and thus confuses two modes of being with each other. Nor does "transcendence" refer to a world outside or beyond this one. Transcendence is always more than the here and now, but is only able to manifest itself from the immanent. It shows itself when objects appear as more than "being present": when human beings experience not only the world but also *that* there is a world, not only objects but also the *being* of objects. This happens, for example, in the classical wonder at the fact that "there is something rather than nothing," or in the existential experience of having been "thrown" into being.

Technology and Mass Existence

Against the background of these basic concepts of Jaspers's philosophy it becomes more clear what is at stake for him in the question of technology. The transformation of society into mass existence threatens the possibility for human beings to exist authentically. The line of reasoning by which Jaspers reaches this conclusion can be summarized as follows: Thanks to the vast expansion of the ability to supply primary life necessities, technology stimulates an enormous population expansion. But this expanded population is now dependent on technology, and could not exist without it. Meanwhile, the technology needed to support the multitude comes to require a comprehensive and efficient organization, whose elements and functions have to be carefully tuned to each other. Society thus acquires the character of an apparatus in which human beings exist not as themselves, but only for their functions. Moreover, technology comes to define not only work but also leisure, which takes shape as the time in which human beings seek through distraction to renew themselves so that they are physically and mentally able to return to work. All of this leads to a leveling of society, according to Jaspers, in which human existence is reduced to mass existence.

How adequate is this analysis? Does technology in fact lead to mass existence? In order to answer these questions, a first essential step is to describe more exactly what Jaspers means by "technology" when he claims that technology threatens human existence. In his outline of technological society Jaspers seems to model his conception of technology on the assembly line, asserting, for instance, that the "joy in work is ruined whenever the working of the universal order is such as to split up the whole into partial functions" (Jaspers 1951, 64) and that "for the many whose work consists of continuous repetitive motions at the conveyor belt," work has become something negative (Jaspers 1953, 110). In light of this conception of technology, the work associated with it can only be conceived as a disconnected fulfilling of a function that leaves no room for personal individuality, while leisure can only be conceived as escapist amusement for the masses that does nothing more than distract them.

Once again, however, this conception was part and parcel of the historical time in which Jaspers wrote. In the 1930s, industry was much less mechanized than it would shortly become; it was a transitional period during which mechanization was expanding, but the image of the traditional workplace was still fresh in mind. The image of workers as no more than extensions of machines, whose behavior was dictated by the equipment with which they worked, stood in such sharp contrast with the traditional workplace that Jaspers and others like him concluded that modern technology alienated human beings from themselves.

The assembly line, however, is of limited value for understanding technology adequately—especially in today's context, when mechanization is less the issue than automation, and the dehumanizing aspects of technology are less evidently present. Once we look beyond this image it is easy to perceive other aspects of technology, including examples of technologies that create new situations that do allow human beings genuinely to "exist as themselves." A variety of new medical technologies, for instance, have opened up all kinds of questions that can legitimately be called existential in nature—such as which treatments to choose, when to stop them, and whether to choose a treatment that may lengthen life but substantially reduce its quality. Technologies such as these create situations that could be called "boundary situations" in Jaspers's terminology. But even in less exceptional situations technology does not stand opposed to the possibility of human beings authentically existing "as themselves." Communication and information technologies such as cell phones and email surely have changed human communication, but have made possible moments of contact between human beings that are genuine and personal, and not merely functional.

In general, one can say that technology has provided us with more rather than less space in which to "be ourselves." With more aspects of our experience falling under our control and influence, ever less is taken for granted—giving rise to an increasing number of situations in which human beings must make personal choices. Indeed, philosophers have even expressed concern that we are faced today with *too many* choices, leading to a state of permanent reflection whose effect can be paralyzing.[8] In short, technological society still provides us with ample opportunity to exist as ourselves, and is not completely stamped with bureaucratic structure.

Here, too, we need to seek the background of Jaspers's alienation thesis in the transcendentalist style of argument that he employs. Here, as in his account of mass production, Jaspers analyzes the relation between technology and mass existence by seeking the *conditions of possibility* for technology in our daily lifeworld. In this case, Jaspers sees two conditions of possibility as playing a role. The first is that, in its present scope, technology could not function without a functionalist and bureaucratic social structure; from the fact that technology requires bureaucracy and functionality, he concludes that contemporary society only admits whatever is functionally and bureaucratically organized. But that is a logical error. Just as the fact that I can only see with glasses does not imply that I can *only* see and cannot hear when I am wearing glasses, so it does not follow from the necessity of a functional and bureaucratic organization for society that society *only* has space for the bureaucratic and functional and none for the personal and existential. However much our society is unmistakably of a bureau-

8. See, for instance, Achterhuis (1993a). The concept of "permanent reflection" comes from Arnold Gehlen.

cratic and functional form, human beings still have the opportunity to exist genuinely as themselves, as can be seen by the examples given above.

The second condition of possibility from which Jaspers understands technology is the neediness of human beings. When technology is understood as that which creates the means for supplying human needs, and when these needs progressively increase, the effort to supply these needs must become so tightly organized that the entire society must become revamped in terms of this required functionality. But it is questionable whether technology can be understood as a response to the task of supplying human needs in this way. By contrast, for instance, the Spanish existential philosopher José Ortega y Gasset, in his *Thoughts on Technology*, defended the view that technology can only be understood in view of its superfluousness. Technology is only possible when human beings suspend their immediate needs, when they delay gratification and seek an indirect, and ultimately more efficient, way to realize them. Technology, for Ortega, therefore is a response less to the need to live than to the desire to live "a good life"; technology is not a matter of the "being" of humans but rather of their "well-being." The organic aspects of human life can be supported in nontechnological ways, but with technology human beings give form to "the good life," a life that is not only biological but also meaningful.[9]

Ortega, like Jaspers, conceives of humans as beings whose existence is not given but must be realized, which they do in their confrontation with the world around them. This world does not immediately offer them everything necessary for "the good life," but neither is it an impregnable fortress. Human beings can transform the world, which they do with the help of technology. Technology thus forms the answer to the tension between the human need for well-being and unruly but pliable reality. In Ortega's view, therefore, technology does not alienate human beings from themselves, but instead offers them the possibility of existing as themselves.[10] Technology makes it possible to do more with life than just survive.

9. Ortega y Gasset (1972, 290–313).

10. In the end, Ortega y Gasset arrived at a negative judgment of the value as well as the technology of the present. Technology, he came to believe, had entered a new phase by severing the connection between design and production. When the work plan and its execution were closely connected, according to Ortega y Gasset, technology had a clearly defined place in work contexts. Once these became disconnected, human beings had at their disposal "technology as such" rather than concrete technologies for specific purposes. They realize *that* they can encroach upon the world in a technological way, and this realization leads to the development of progressively more technology. The problem, for Ortega y Gasset, is that technology cannot dictate its own applications; human beings have to. They suddenly acquire such an enormous ability to design and build things that they lose sight of what to do with this ability. "To be a technician and only a technician means to have the possibility to be everything and consequently to be nothing. Precisely because technology is laden

But to recognize such aspects of technology a merely transcendental approach is insufficient. For the issue is not simply the conditions of possibility for technology, but also which possibilities, if any, are the new ones *to which it gives rise.* Technology is unquestionably necessary in order to supply basic human needs, but cannot be reduced to this task. If, however, technology is "forwardly" approached in terms of the concrete roles it plays in human existence, this directs our attention to the existential possibilities that technologies open up for us, rather than close off. Instead of merely threatening human existence, technology then appears to create new ways of existing. An existential philosophy of technology needs to explore how technology opens up these new possibilities through which human beings can realize their existence and to examine how this happens.

THE NEUTRALITY OF TECHNOLOGY

Jaspers's philosophy of technology changed course after World War II, and has a completely different flavor in *The Origin and Goal of History* (1953) and *The Atom Bomb and the Future of Man* (1958). Elisabeth Hybašek has pointed out that this development takes place in the context of Jaspers's new view of "the masses," which he no longer regarded merely negatively (Hybašek 1984, 263–65). Instead of viewing technology simply as a threat to it, Jaspers's revised view posits technology as what is *at stake* in authentic human existence. To overcome "the demonism of technology" requires us to realize that technology is ultimately nothing more than a collection of means, neutral in themselves, for ends that we ourselves set. Humanity needs to pose and answer for itself the question of what it wants to do with technology. But this new approach, too, is not free of problems.

Limits of Technology

Jaspers first set forth his new approach in *The Origin and Goal of History,* where he analyzes modern technology and its role in human existence and social life. He defines technology much more positively and clearly as "the procedure by which scientific man masters nature for the purpose of molding his existence,

with possibilities it is merely empty form—like the most formalistic logic—and not in a position to determine the content of life. For this reason our modern era, which is the most intense technological period which human history has ever known, is also the emptiest" (Ortega y Gasset 1972, 151). I do not share this judgment, as will shortly become clear, though I do find fruitful his observation that technology does more than satisfy vital human needs.

delivering himself from want, and giving his environment the form that appeals to him" (Jaspers 1953, 98). His approach consists now of investigating its limits: "The appraisal of technology depends upon what is expected of it. A clear appraisal presupposes clarity concerning the limits of technology" (118). Jaspers then adds, "The limits of technology lie in those presuppositions of all technological realizations which can never be overcome" (118), and which are not susceptible to technological control.

The first limit that Jaspers mentions is the most important: technology, according to him, is limited by the fact that it is a means and requires direction. With this claim Jaspers sets himself off from two other conceptions of technology that he deems inadequate. The first *glorifies* it, seeing technology as "an all-embracing realization of the allegedly true and legitimate environment of man" (114), as a decisive liberation from nature. The second conception *despises* technology, and while diametrically opposed to the first likewise captures only part of the truth. In this conception technological control fails to liberate us from nature but instead destroys nature and humanity along with it. Jaspers considers both positions—including the second, which, remarkably enough, was his position in *Man in the Modern Age*—to be untenable. Technology itself follows no particular direction, neither toward a completion nor toward destruction. Only human beings can give it direction; it is in itself neutral, and it requires guidance. It is in no position to give itself ends and is only the means for realizing ends provided by human beings. Technology now appears as a task or challenge for human beings, calling for them to ask to which ends they want to apply it, and which not.

Jaspers now denies the conception implicit in his earlier position that technology has an intrinsically destructive effect. Still, the claim that technology is ultimately a neutral means amounts to a nuancing, rather than an outright repudiation, of his earlier diagnosis of its "demonism." For Jaspers maintains his conviction that technology at present acts like a demon that threatens authentic human existence—but does not now view the demonism as an intrinsic property of technology. Rather, technology has *become* demonic thanks to the manner in which human beings have handled it. While in itself technology is no more than a means, humans have interpreted it as an end in itself, allowing it to function as an independent and menacing power while not actually being so itself. The task now for human beings, according to Jaspers, is to treat it once again as a means and reassert sovereignty over it.

The other limits of technology that Jaspers mentions, besides that it is a means and requires direction, are of lesser importance for understanding the existential perspective, though for the sake of completeness I shall mention the most important without critiquing them all at length. Technology, according to Jaspers, is restricted to the mechanical. Those who speak of the supposedly limitless possibilities opened up by technology fail to appreciate that technology only produces what can be attached to the skeleton provided by its mechanism. "The steps to be taken with regard to nature in cultivation and breeding, with regard to man in education and communication, the creation of the works of the spirit, indeed invention itself, cannot be performed according to the rules of technique" (119). Jaspers also finds that technology is restricted to the *lifeless.* Given that technology can only deal with the mechanical, it can only deal with living things as if they were dead. While the traditional methods of flower cultivation produce blossoms as if they were art works, modern technologized plant cultivation turns out flowers the way the automobile industry turns out car models. Technology, Jaspers says, is also restricted to the *universal.* By this he means that technology is always transferable and accessible to all cultures, and therefore necessarily impersonal. Finally, technology is *bound to substances and forces that are limited*—the quantity of raw materials and energy sources is finite—and bound as well to *human beings,* whose labor keeps it going (120–21).

Jaspers's new perspective allows him to discern not only a threatening side of technology but also a more positive side, one that presents human beings with new possibilities of existence. He describes these new possibilities of existence as involving a new closeness to nature (116), several aspects of which he discusses. One is the way in which the use of scientific devices brings us into contact with and opens up access to nature as revealed by physics. When human beings are open to it, the functioning and fabrication of technical devices gives them a proximity to the laws behind nature. Another possibility involves what Jaspers calls the *beauty of technological constructs.* This beauty does not flow from their functionality or how they are decorated, but rather from the seamless ways that they can be integrated in human life, and from the experience that "there are solutions inherent in the thing itself, which are found, as it were, through the striving after eternal, predetermined forms" (117). Yet another possibility for a new closeness to nature involves the *extension of perception.* The microscope, telescope, media, imaging devices, and transportation methods open up new worlds that otherwise would remain closed off. The worldwide media and transportation networks allow human beings to experience the entire globe and have given rise to, finally, a *new consciousness of the world,* a world that is a "closed whole." Jaspers concludes:

> Thus the technological world contains humanity's new possibilities, the specific pleasures of technology, the achievement of technology in

> extending the experience of the world, in rendering the whole planet and all the elements of existence present to concrete experience; the foundations are laid for a playful mastery of matter that will lead to pure experiences of the sublime. At the present time, however, all this is still a rare exception. Apart from skill, this new nearness to nature and to all things postulates the sovereignty of man, who, penetrating with his power of apprehension into the realm of that which is alien to nature, creates presentness out of the whole, which is not immediately existent. . . . Much easier is silting up in thoughtless, empty and mechanical functioning, alienation in automatism, losing oneself in diversions, growing unconsciousness, the residue of nervous excitation. (118)

Guiding Technology

Jaspers's later approach to technology also contains a new perspective on the relation between human beings and technology. While in *Man and the Modern Age* he concluded that technology poses to human beings the task of retrieving their authenticity, he came to see retrieving technology—bringing it back under their control—as a task in its own right. While technology is ultimately a neutral means for the realization of human ends, according to Jaspers, things have gone awry and technology has become itself the goal. Technological development races forward unchecked, without human beings bothering to inquire into the ends to which it might be applied. Everything that can be technologically done, is. This situation poses the task of taking back control over technology, using it once again as a means, and consciously deliberating about the ends to which it is put.

While Jaspers presented these thoughts initially in *The Origin and Goal of History,* he elaborated them further in *The Atom Bomb and the Future of Man.* There he drew attention to the atomic bomb as a consequence of the unchecked progress of technology that forced on human beings the inescapable realization that technology is in need of control. But to control technology, according to Jaspers, requires in turn a reversal in thinking, a "revolution of thought" in which technological thought, or the "intellect" (*Verstand*), is transformed into a "new way of thinking," an "existential" way of thinking that he calls "reason" (*Vernunft*), in which individuals exist authentically as themselves (Jaspers 1963, 209). When humans only think intellectually, they only solve technological problems, leaving their real problems unaffected. Only an authentic way of thinking in which individuals exist as themselves will allow them to turn the situation in which they find themselves into *their* situation, for which they are *responsible.* "What grows out of the free acts of countless men and comes upon us like an overwhelming tide of events is no mere tide of events. Every individual acts in it as a free agent. However powerless he may feel, no one is wholly powerless. . . . However minute a quantity the individual may be among the factors that make history, he is a factor. He cannot attribute it all to a tide of events of which none is his doing" (7–8). The question that technology provokes for us is thus no longer, "How can we free human beings from the grip of technology," but rather, "What shall we do with technology?" The predicament that the atomic bomb has thrust upon humanity did not arise of itself, nor was it caused solely by technology—and thus it cannot be changed either by altering itself or technology only. Human beings brought about this situation, and only they can change it. But the change cannot be effected by technological/intellectual thinking, which can plan, organize, create, design, and regulate, for this thinking presupposes preexisting goals. The intellectual, like the technological, is only a means and not in a position to provide its own goals, which only free, individual human beings can provide. According to Jaspers, a transformation of the contemporary situation can only be brought about by a reversal of thinking, in which intellectual thinking is transformed into reason—a way of thinking in which human beings are actually thinking *as themselves,* and guide the development and use of technology on the basis and in view of the human possibility to exist as oneself.

Reason, however, does not provide any specific principles or prescriptions for guiding technology. Those who seek such principles are not thinking authentically if they try to hold themselves to "ethical standards" and thereby abstain from being "as themselves." Reason will never provide a simple solution or definitive answer to the question, "What is to be done?" It is a way of thinking through and by which humans engage with the world as themselves, and from out of such engagement seek to respond to the situation in which they find themselves. A turn toward reason therefore entails, for Jaspers, a recovery of a sense of *responsibility* for technology. The intellect is only a means to an end, and with respect to their ability to use this intellect humans are interchangeable. In the use of reason, by contrast, individuals set their own goals, which first opens the possibility of their being truly shocked by the threat of annihilation that the atomic bomb poses to humanity and life on earth. They experience themselves as coresponsible, because failure to act is also an action, thus a choice. Only such thinking can turn the contemporary predicament into a task and allow humanity to seek new goals for applying technology.

Beyond Alienation?

At the beginning of this chapter, I criticized Jaspers's early approach on the grounds that he fails to do justice to the role that technology plays concretely in human existence. To what extent, then, is this problem truly overcome in his later conception of technology? How convincing is the alternative that Jaspers proposes to the classical diagnosis of alienation?

The view that technology is to be understood as a neutral means is highly unusual in the philosophy of technology. Customarily, a philosophy of technology begins by emphasizing that technology is *not* neutral, noting that technologies do much more than simply achieve the goals for which they were instituted. The new possibilities that technologies open up always change the context in which they play a role. An automobile, for example, is a means for transporting us from point A to point B—who can deny that?—but meanwhile does much more as well. First, technologies reshape the very ends that we use them to reach. An automobile allows us to travel longer distances in shorter periods, and therefore makes it possible to live at greater distance from work, thus contributing to a greater separation of the sphere of labor and the sphere of leisure. Means and ends are internally, not externally, linked here, for the goal would have been different had a different means been employed. Second, a technology does much more than realize the goal toward which it is put; it always helps to shape the context in which it functions, altering the actions of human beings and the relation between them and their environment. Without the automobile, for instance, our everyday social ties would be restricted to a much smaller geographical area. Furthermore, automobile drivers have a much different relation to their surroundings than do, say, pedestrians and bicyclists.

Jaspers, however, is not speaking about the neutrality of concrete technologies but rather of "technology" itself. He is not addressing the question of the connection between specific technologies and their ends, but the question of why human beings have "technology" in the first place. Technology is neutral, according to Jaspers, in the sense that technological development itself cannot be guided in a particular direction. The development of new technology requires, according to him, that a clear purpose be in place. The neutrality of concrete technological artifacts would make human beings sovereign only over those artifacts—but Jaspers goes much further, and claims that human sovereignty extends to technology itself. Precisely this idea of human sovereignty over technology is what makes his position problematic. It implies that technology and culture are two independent domains and fails to recognize their factual and inescapable entanglement. Human beings develop technology always from within a context which is coformed by technology: technology has helped to evoke and shape their needs and desires, likes and dislikes. Without the possibility of abortion there would be no demand for prenatal screening, while without computers there would be no need for email. Technologies coshape the human world and thus also human relations with technology itself. Human beings are not sovereign with respect to technology, but are, rather, inextricably interwoven with it.

That Jaspers fails to see this interweaving has to do, once again, with his transcendentalist approach. His ideas about the neutrality of technology and the necessity of human sovereignty over it are based on an analysis of "the limits of technology." He again attempts to understand technology in terms of its conditions of possibility, since the limits he discerns are in fact a kind of "negative" condition of possibility. It is plausible to assume that technology can only be developed when its creators have concrete goals in mind. But the "limits" should not be mistaken for technology itself. The fact that technological development needs to be understood through the intentions of its creators does not imply that the role technology goes on to play can be completely reduced to these intentions.

The device that we now call the telephone, for instance, was initially intended as a contrivance to assist the hard of hearing, a sort of precursor to the hearing aid, which would amplify the voice of the interlocutor (Ihde 1993a, 116). The context in which this device came to function redefined it as what we now know as the telephone; furthermore, this new definition—which its developers did not have in mind—in turn reshaped many aspects of daily life in ways that would not have occurred without the telephone. This inextricable connection between human beings and technology, however, remains invisible when technology is exclusively conceived from the point of view of its "limits." Such an approach only makes visible the *conditions* of technology—in this case, cultural conditions—and not the role that it plays concretely in society and human existence.

The transcendentalism of Jaspers's later approach to technology thus leads, not to a diagnosis of alienation, but to its mirror image—to the thought that human beings are sovereign with respect to technology conceived as neutral in itself. Both approaches take as their point of departure an untenable separation between human beings and technology. In the one case technology has human beings in its power, while in the other humans have power over technology. This backward thinking, therefore, not only makes him confuse technology itself with its conditions of possibility, but also makes him fail to recognize that human beings and technology are inextricably interwoven. It is, instead, precisely on the basis of this interwovenness that technology serves to coshape that from which the existential perspective of the philosophy of technology takes its point of departure: human existence.

CONCLUSION

Well before the philosophy of technology established itself as a philosophic discipline, Karl Jaspers formulated a systematic existential approach to technology. In his approach, a negative judgment predominates: technology leads to a system of mass production, which suffocates the human possibility of existing authentically and alienates human beings from their material environment. The later Jaspers saw this "demonism of technology" as a task to be confronted rather than as an unavoidable destiny. When and if human beings can turn away from intellectual thinking to another kind of thinking wherein they can exist authentically, and when and if they see that technology is ultimately a neutral means for realizing goals that they themselves set, will they at last recover sovereignty over technology and be in a position to guide it.

A closer analysis of Jaspers's position reveals that the path by which he has arrived at his diagnosis is not unproblematic. My main critique of his analysis is that he understands technology principally in terms of its conditions of possibility. When he inquires into the relation between human beings and technological production, he does so with an eye to how technology emerges from a system of mass production. Similarly, when he inquires into the role of technology in human existence, he reduces technology to the bureaucratic and functional organization of society from which it arises. Finally, when Jaspers asks how the threat he perceives technology as posing can be overcome, he reduces it to its "limits," the most important of which is that it is merely a means that is unable to give direction to itself.

Such an approach is doomed to produce a distorted image of technology. In place of technology, Jaspers uncovers only its conditions of possibility. Technology then seems to create only a functional world and to put in place obstacles that prevent human beings from authentically realizing their existence. A broader approach is required to do justice to the role of technology in human existence, one that elucidates the role of concrete technologies without seeking to reduce them to their underlying conditions. It should be at least intuitively clear by now that such an approach produces a much different picture. The most important, and misguided, consequence of Jaspers's transcendentalist approach is the isolation of technology and the human beings in whose existence it plays a role into two separate spheres. Whether technology is seen as a threat to human existence or as what is at stake in human existence, in each case the two spheres are viewed as externally related rather than as internally interwoven and interdependent. Human beings, however, are not sovereign with respect to technology, for technology itself coshapes the ways in which they relate to it. Nor are they only threatened by technology, for technology also coshapes new existential dimensions of their existence. Precisely because it gives form to human existence, this mutual interweaving of human beings and technology needs to become the point of departure for an existential analysis of technology.

This critique of Jaspers's philosophy of technology, however, does not mean that we should write off his analysis completely. His inquiry into the role that technology plays in the way human beings realize their existence represents an important and undervalued perspective in the philosophy of technology, and we must not lose sight of it. We have to pursue it, however, through a closer examination that does more justice to the role technology concretely plays in our daily lives. In the second part of this book I try to formulate alternate answers to the existential questions with regard to technology that Jaspers has put on the agenda. In this alternate approach, I do not approach technology in terms of its conditions of possibility, but rather as a set of concrete technological artifacts that play an active role in human existence. The intuitions expressed in this chapter about this existential role of artifacts can then be more systematically elaborated.

Jaspers's philosophy of technology, to conclude, contains a remarkable lacuna. Jaspers discerns three modes of being: Dasein, existence, and transcendence. In his philosophy of technology, he elaborates what he views as the implications of technology for Dasein (i.e., the coming into being of mass products) and existence (i.e., the rise of mass culture), but he fails to address transcendence. This may be connected with the fact that transcendence, albeit framed and elaborated in a different vocabulary, figured so prominently in the thought of his contemporary, Martin Heidegger. His hermeneutical approach is the subject of the next chapter.

2

The Thing About Technology

INTRODUCTION

"The thing things."[1] When philosophers want a good laugh, they often trot out a quote from Heidegger such as this one, conjoining a noun with a verb made from it. In this chapter, however, I shall argue that the thought expressed by the phrase "the thing things" is extremely important for the philosophy of technology—though I shall be developing it in a different way than Heidegger himself did. Heidegger's phrase expresses the thought that what a thing "does" has to be understood from out of the thing itself, without being reduced to something non-thingly. The importance of such an approach appeared in the previous chapter, where it became clear that technology cannot be understood solely in terms of its conditions of possibility, but should also be analyzed in terms of concrete artifacts that play a role in the relation between human beings and their world.

Heidegger's philosophy offers, however, only few apparent points of contact with such an approach. I will make clear that the very thinker who asserted that "the thing things" formulated a philosophy of technology that contains the same flaw that appeared in the previous chapter;

1. Heidegger (1971c, 174).

like Jaspers, Heidegger thinks "backward," and reduces technology to what underlies it. In the philosophy of technology, Heidegger represents, par excellence, the classical perspective that has become so contested in the past few decades. His philosophy is complementary to that of Jaspers within the classical phenomenological approach to technology. While Jaspers raises the existential issue of the role technology plays in human existence, Heidegger raises the hermeneutical issue of the role technology plays in the way human beings encounter and interpret reality.

Here, too, a critical analysis will be required in order to make Heidegger's hermeneutic philosophy of technology relevant to a "turn toward the artifact" in the philosophy of technology. This is my aim in the present chapter. I shall first outline the main themes of his philosophy of technology. In doing so, I shall aim not at completeness—good introductions to Heidegger's philosophy of technology abound[2]—but at an explanation that will lay the groundwork for the next paragraphs, in which I shall critically analyze Heidegger's position. I shall initially focus on "The Question Concerning Technology" and "The Memorial Address"—the two texts most commonly cited in dealing with Heidegger's philosophy of technology, and the ones that discuss the issue most explicitly—treating for my purposes the latter essay as essentially an epilogue to the former. Then I shall critically analyze his approach and the consequences that flow from it for his picture of technology.

In the process I shall broach a number of accusations that are regularly made in connection with Heidegger's philosophy, such as that it is "monolithic," "abstract," and "nostalgic." Heidegger is often charged with having had too little contact with concrete technological practices and with wrongly opposing himself unilaterally to the changes wrought by technology. Although to a large extent I share these views, I find that they remain too superficial to adequately criticize Heidegger.[3] They remain too external to Heidegger's analysis and therefore are in their own specific ways too monolithic and abstract as well. I will, therefore, elaborate these forms of criticism and relate them to the transcendental manner of thinking that I have already described. This will make clear that Heidegger can support his nostalgic preference for traditional technology only by selectively drawing from two different—and incompatible—approaches to technology.

My critique of Heidegger reveals the necessity for another kind of thinking about technology, one that takes the thought that "things thing" more seriously than Heidegger will appear to do—at least in his later period, for I find that his

2. These include Borgmann (1987); Ihde (1993a); and Zimmerman (1990).

3. An earlier, though much less fully worked out, version of this critique appears in Verbeek (2000, 281–99).

earlier work, prior to "The Question Concerning Technology," provides just the right points of departure for this other kind of thinking that I am after. In order to estimate the value of these points of departure, in the last part of this chapter I shall analyze the development of Heidegger's "thinking about things," drawing on my reading of his early principal work, *Being and Time,* and other texts up to "The Memorial Address." My examination of the course of Heidegger's approach to artifacts reveals how a transcendental manner of thinking gradually crept into his philosophy, culminating during the period of its so-called turn or *Kehre.* But exposing this transcendentalism, in turn, makes it possible to see how it can be avoided. This will create the possibility of seeking new answers to the questions, of undiminished importance, that Heidegger poses about technology: What is the significance of technology for the way human beings encounter reality; and how does it affect the manner in which they interpret the world?

HEIDEGGER'S PHILOSOPHY OF TECHNOLOGY

"But we are delivered over to [technology] in the worst possible way when we regard it as something neutral; for this conception of it, to which today we particularly like to do homage, makes us utterly blind to the essence of technology" (Heidegger 1977a, 4). We can only surmise whether this remark, drawn from the first page of "The Question Concerning Technology," was targeted directly at Jaspers. Whether it was or not, it reveals straightaway that Heidegger's philosophy of technology is of a completely different character than the one outlined in the previous chapter. Heidegger is adamantly opposed to the contention that technology is "a means to an end" or "a human activity." These two approaches, which Heidegger calls, respectively, the "instrumental" and "anthropological" definitions, are indeed "correct," but do not go deep enough; as he says, they are not yet "true." Unquestionably, Heidegger points out, technological objects are means for ends, and are built and operated by human beings, but the essence of technology is something else entirely. Just as the essence of a tree is not itself a tree, Heidegger notes, so the essence of technology is not anything technological. Even so, he gives the concept "essence" an unconventional meaning, as will shortly become clear.

What, then, is technology, if it is neither a means to an end nor a human activity? I shall elaborate Heidegger's answer to this question in three stages. First, I shall explain why technology, for Heidegger, cannot be thought of as a means; second, I shall explain why it also cannot be conceived as a human activity; third and finally, I shall discuss Heidegger's contention that technology must be understood as "the greatest danger."

Technology as "Revealing"

Technology, according to Heidegger, is much more than only a means to an end; it must be understood as "a way of revealing" (12). "Revealing" is one of the terms Heidegger adopted in order to make it possible to think what, according to him, is not thought anymore. It is his translation of the Greek word *alètheuein,* which means to dis-cover, or draw out of concealment—to uncover what was covered over. Related to this verb is the independent noun *alètheia,* usually translated as "truth," though Heidegger insists that a more adequate translation would be "un-concealment."

What does Heidegger mean when he says that technology is "a way of revealing"? A brief excursion into his ontology is needed in order to answer this question. What we call "reality," according to Heidegger, is not present the same way in all times and all cultures (Seubold 1986, 35–36). "Reality" is not something absolute that human beings can ever know once and for all; it is relative in the most literal sense of the word—it exists only in relations. Reality in itself is inaccessible for human beings. As soon as we perceive or try to understand it, it is not "reality in itself" anymore, but "reality for us." Only in the relation of human beings to "that" with which they are related does "that" become reality, as I explained in the introduction.

For this reason, Heidegger calls "that" "the concealed." He has no better name for it, for each more specific name would require a relationship with it, which would already disclose it in a specific way. Only in the relationship humans have with entities do these entities become reality for them—they "emerge out of concealment into unconcealment," in Heidegger's words. This emerging out of concealment into unconcealment is what Heidegger names with the term "revealing." Revealing therefore means the entering into a particular relation with reality in which reality manifests itself in a specific way. It is in and through revealing that reality comes to presence for human beings. Reality is always already marked by the relation that humans have to it.

The revealing of reality, however, cannot happen in any arbitrary way. Something else precedes and determines it, which Heidegger calls the "way of unconcealment" that holds sway in a particular epoch. There can only be revealing when there already exists a particular understanding of what "being unconcealed" means. In other words, in revealing, reality comes to be, and that presupposes an understanding of what "being" means. The manner in which reality can come to be (or in which unconcealment can happen) depends on what is understood by "being": what it means "to be," or, in Heidegger's vocabulary, the "manner of unconcealment that holds sway." "Wherever man opens his eyes and ears, unlocks his heart, and gives himself over to meditating and striving, shaping and working, entreating and thanking, he finds himself everywhere already brought into the unconcealed. The unconcealment of the unconcealed has already come to pass whenever it calls man forth into the modes of revealing allotted to him" (Heidegger 1977a, 18–19).

The relation between humans and being is the linchpin of Heidegger's entire thought. The question of being appears trivial: What is new in the claim that reality must "be" in order that human beings can relate to it? Heidegger agrees that the question is trivial—even so trivial that human beings have come to overlook it, for Western thinking has concentrated on the question of "what" a thing is to the exclusion of "that" it is. But when this question is thought afresh, it is not as trivial as it might appear—and it becomes relevant to technology. According to Heidegger, an important aspect of being is that it is never fixed for all time, but changes throughout history. In different epochs humans have understood it differently. The way of revealing of modern technology rests upon the most recent understanding of being—which Heidegger actually views as a lack of understanding, for he sees the course of Western history ever since the ancient Greeks as involving the progressive forgetting of being, a forgetting that has culminated in modern technology.

The ancient Greeks, before Socrates, had not yet forgotten being, as the word *alètheia* testifies. The meaning of "unconcealedness" is still heard in it; beings are conceived as entities that must be wrested from concealment, that have come into presence. But the way in which being was understood changed, for it soon came to be understood as "a being," or "an entity." Thus the fundamental distinction (the "ontological difference") that Heidegger draws between being and beings disappeared and was eventually forgotten.

The forgetting of being, according to Heidegger, began with Plato. While for the Presocratics "being" still meant "emerging out of concealment into unconcealment," for Plato it began to mean "essence." "Being" meant embodying an idea, which forms the "essence" of the entity. In the Middle Ages the forgetting of being took a new path. After Greek and Christian thought intersected, "being" began to mean "shaped by God." "Coming to be" was then no longer conceived as an emerging out of concealment into unconcealment, but as an act of creation carried out by God—"being" was understood as the effect of a cause rather than as the "happening" of the transition from concealment into unconcealment. God as the ground of all beings came to be understood as a being Himself—a fatal confusion, according to Heidegger, even when God is conceived as the *highest* being, the *ens summum.*

At the beginning of modern philosophy René Descartes moved further still, regarding this appeal to an extra-mundane ground as superfluous: "being" for him meant to be an object for a subject, *res extensa* as opposed to *res cogitans.* The capstone of the forgetting of being, as far as philosophy goes, was set into place by Friedrich Nietzsche, in whose work "being" means merely "being usable for the Will to Power." This last meaning of being, according to Heidegger, finds its material realization in modern technology. Being comes to mean: available for production and manipulation, raw material, "standing-reserve."[4]

Thus when Heidegger claims that technology is a way of revealing, he means that it involves a specific kind of being. It rests upon a relation between human beings and reality, one involving a way of revealing in which human beings "set upon" and "challenge" what they bring forth as real. As a result, reality is understood in terms of what is available to and can be controlled by human beings. The way of revealing of the ancient Greeks provides a ready contrast. The word "technology," Heidegger points out, comes from the ancient Greek word *technikon,* which refers to everything pertaining to *technè. Technè* comprised both craft and art alike. It was a form of *poièsis,* which is generally translated loosely as "making." But Heidegger proposes what he claims is a more adequate translation: "bringing-forth" (*Her-vor-bringen*). This, he says, preserves a meaning expressed in the word *alètheia:* the emerging out of concealment into unconcealment, or the letting come to presence. "Technology is a mode of *alètheuein.* It reveals whatever does not bring itself forth and does not yet lie here before us" (13). Technology as *technè* needs to be understood, according to Heidegger, as a manner of revealing involving a transition from concealment to unconcealment, and he elaborates this point with reference to the famous Aristotelian doctrine of causality.

Aristotle distinguished between four "causes," or ways in which to speak about what is responsible for something to be brought into being: the *causa efficiens* (efficient cause), the *causa materialis* (material cause), the *causa formalis* (formal cause), and the *causa finalis* (final cause, or the end to which the produced object is put). These four causes are coresponsible for the produced object; together they make it what it is. But the word "cause," according to Heidegger, does not

4. For an overview of Heidegger's history of being in connection with modern technology, see the chapter titled "The History of Productionist Metaphysics" in Zimmerman (1990, 166–90).

characterize well what is happening here. The word is a Latin translation of the Greek *aition,* meaning "that to which something is indebted." Aristotle's causes, then, must not be conceived to mean that which brings about or effects, but as that which is coresponsible for what is brought forward. The phrase "coresponsible for" is meant to signal that "bringing forward" or "pro-ducing" is not simply a "result" of the four causes—and certainly not of only the *causa efficiens*—but that it is instead a happening to which these causes *contribute.* For Heidegger, this happening or event is the "coming to appearance" or the emerging out of concealment to unconcealment.

The ancient Greeks, therefore, viewed *technè,* according to Heidegger, as a manner of disclosing reality in which the presencing process was not brought about by human beings alone, but something in which—in Jaspers's terms—a certain transcendence played a role. The presencing process, that is, cannot be reduced without remainder to human "causes," but is also indebted to something "over and beyond" as well. Human activity, together with the eventual form of what is brought forward, the material from which it is fashioned, and the goal to which it is to be put, forms the *occasion* for what is to come forward. The four *aitia* are ways in which a being is, as it were, "helped" to come into being. How, then, is this related to the revealing of modern technology?

> It, too, is a revealing. Only when we allow our attention to rest on this fundamental characteristic does that which is new in modern technology show itself to us. And yet, the revealing that holds sway throughout modern technology does not unfold into a bringing-forth in the sense of *poièsis.* The revealing that rules in modern technology is a challenging (*Herausfordern*), which puts to nature the unreasonable demand that it supply energy which can be extracted and stored as such. . . . The revealing that rules throughout modern technology has the character of a setting-upon, in the sense of a challenging-forth. Such challenging happens in that the energy concealed in nature is unlocked, what is unlocked is transformed, what is transformed is stored up, what is stored up is in turn distributed, and what is distributed is switched about ever anew. . . . What kind of unconcealment is it, then, that is peculiar to that which results from this setting-upon that challenges? Everywhere everything is ordered to stand by, to be immediately on hand, indeed to stand there just so that it may be on call for a further ordering. Whatever is ordered about in this way has its own standing. We call it the standing-reserve (*Bestand*). (16–17)

Modern technology, according to Heidegger, thus involves a much different way of disclosure than the *techné* of the Greeks. It is no longer a bringing-forward but a setting-upon, a challenging and ordering. Modern technology approaches reality as raw material that is on call to be put to use. In the modern technological approach to reality, humans no longer encounter entities that "emerge from concealment into unconcealment." Reality acquires its identity from what can be done with it. Being, conceived as the event in which the unconcealed emerges from concealment, is thereby forgotten. There is no more "concealment," no "beyond" that plays a role in the bringing-forward of things; meaning is all a function of human doing. Within the ambit of modern technology, what counts as real is only what is makeable and controllable rather than what escapes such making and controlling, what transcends human beings. Reality appears within the scope of modern technology as "standing-reserve" (*Bestand*), as the storehouse of available raw materials.

Technology as "Enframing"

This explanation, then, shows why technology for Heidegger must not be conceived as a means to an end. Heidegger's next move is to show that this technological way of revealing is not a human activity. As he writes in "The Question Concerning Technology":

> Who accomplishes the challenging setting-upon through which what we call the actual is revealed as standing-reserve? Obviously, man. To what extent is man capable of such a revealing? Man can indeed conceive, fashion, and carry through this or that in one way or another. But man does not have control over unconcealment itself, in which at any given time the actual shows itself or withdraws. . . . Since man drives technology forward, he takes part in ordering as a way of revealing. But the unconcealment itself, within which ordering unfolds, is never a human handiwork. (18)

As became clear above, human beings only gain access to reality thanks to the meaning that being has in the epoch in which they live. If disclosing reality is a "letting come to be," this implies that it is dependent on what it means "to be." Every way of "revealing" presupposes an understanding of what it means to be "revealed" or "unconcealed." The technological way of revealing, according to Heidegger, presupposes an understanding of being that makes humans "order the real as standing-reserve" (19). The specific form of unconcealment behind this, which challenges humans to disclose beings as raw material, Heidegger names the *Gestell* (sometimes translated, neologistically, as the "enframing"). The German prefix "Ge-" indicates that the word is a collective noun meant to name the gathering together of what is ordered. In German, for instance, a mountain range, or gathering together of a set of peaks, is called a *Gebirg;* analogously, by the *Gestell* or enframing Heidegger means to refer to what collects the setting-upon (*stellen*) into a whole.

The *Gestell* is the form of unconcealment, the "way of being" of reality, that holds sway in the epoch of technology. This form of unconcealment lets come to presence whatever is revealed as standing-reserve. As Otto Pöggeler puts it, "Where beings are mere reserve and where revealing is a setting upon and ordering of the reserve, there unconcealedness holds sway as enframing (*Gestell*). . . . In the enframing of technology the Being of beings is presentability and deliverability, the disposability of the reserve; what presences in Being, the respective occurring of the unconcealedness which is not at one's disposal, remains forgotten" (Pöggeler 1991, 198).

In the *Gestell,* according to Heidegger, the essence of technology is to be found. "Essence" (*Wesen*), however, is to be understood here not as a noun but as a verb: it is the way of being that holds sway in modern technology. Technology is present for humans as the *Gestell* that lets them reveal reality as standing-reserve.

Where does the *Gestell* come from then, if unconcealment is not effected by human beings? It is being itself, Heidegger says, that shows itself in the form of the *Gestell.* The *Gestell,* he writes, is a *Ge-Schick* or destining (Heidegger 1977a, 24), derived from the German word *schicken,* meaning "to send." *Ge-Schick* is the collective name for what sends, what starts human beings off on the way to a particular way of revealing. But it also means destiny, or what is given to human beings, or what is granted them. What one hears in the word is that the dominant way of unconcealment is outside the power of human beings to alter; they always "come upon" a particular form of unconcealment. Reality always already "is" in a certain way when human beings enter a relation with it—"being" always already has a meaning for them. The "meaning of being" is the ground of their entire relation to reality, for it determines how reality "is" for them.

That the way of revealing of our epoch has the character of a "challenging," according to Heidegger, is thus a consequence of the destining of being. But the "coming to appearance" of the *poièsis* of the ancient Greeks, in his view, was also a *Geschick.* He has no better words for it: no further questioning is possible into the way in which being manifests itself. It is not the result of human doing, for this doing always already takes place in the context of a particular meaning of being. One can only say that being has this meaning now, and that there is no use trying to find a cause for it: it is a "sending," a destiny that actively announces itself to us.

Technology as "the Greatest Danger"

In "The Question Concerning Technology," Heidegger not only discusses the issue of how modern technology must be understood, but concludes with an evaluation of it: the *Gestell,* he claims, is to be seen as a danger, even as "the greatest danger." But what is so dangerous about technology?

Each destining of being, each form of unconcealment, according to Heidegger, harbors within it the danger that humans will misinterpret the unconcealed; specifically, that it will not be thought in its relation to being. He says, for instance, that in an epoch when everything presences itself in a cause-and-effect structure, the ground of all beings can be conceived as a God as the ultimate cause of all things, as was the case in the Middle Ages. When this happens, "the essential origin of this causality" remains hidden (26): in other words, the process by which things emerge from concealment, which first makes it possible for beings to be understood in terms of cause and effect at all, remains hidden. This possibility of misinterpretation is what Heidegger calls "the danger." It is also present in the *Gestell:* "Since destining at any given time starts man on a way of revealing, man, thus under way, is continually approaching the brink of the possibility of pursuing and pushing forward nothing but what is revealed in ordering, and of deriving all his standards on this basis. Through this the other possibility is blocked, that man might be admitted more and sooner and ever more primally to the essence of that which is unconcealed and to its unconcealment, in order that he might experience as his essence his needed belonging to revealing" (26).

But Heidegger calls the danger harbored by the *Gestell* "the greatest danger." He gives two reasons for this. First, it threatens humanity himself, for "he comes to the very brink of a precipitous fall; that is, he comes to the point where he himself will have to be taken as standing-reserve" (27). When human beings take themselves as standing-reserve, they lose not only themselves but also the ability to be open for being, for the "happening" of "coming into being." Second, when the *Gestell* holds sway, it "drives out every other possibility of revealing" (27). The danger of the *Gestell* is thus not only that the event of being is *neglected;* in addition, because the *Gestell* reduces everything to human domination and control, it becomes no longer even *possible* to think about being and the process of emergence from unconcealment. Not only is it overlooked, but it can no longer even be *seen*. For when human beings would strive deliberately to disclose reality in a different way, they only reconfirm the power of the *Gestell*. Humans then aim to dominate and control even the meaning of being that is the ground for their relationship with beings, and thus they are still caught up in the will to power of the *Gestell*. The meaning of *technè* as bringing forward from concealment into unconcealment, which the ancient Greeks saw in *poièsis*, now belongs definitively to the past. In the *Gestell*, it is not so much that unconcealment is misconstrued, it is that unconcealment cannot be thought at all. The rule of the *Gestell* "threatens man with the possibility that it could be denied to him to enter into a more original revealing and hence to experience the call of a more primal truth" (8).

Nevertheless, Heidegger saw a way out of the danger. His argument takes an almost dialectical turn when he claims that the danger harbors in itself also a saving power, and quotes the following lines from Friedrich Hölderlin: "Wo aber Gefahr ist, wächst / Das Rettende auch" [But where danger is, grows / The saving power also] (28). When human beings see and think the danger, they thereby at the same time open for themselves the possibility of a way out. By resisting the temptation to see technology as a collection of artifacts that serves as means for ends and to see it as a human activity, human beings can put themselves in a position to think about the essence of technology, and in doing so to allow the dimension of unconcealment hidden in the *Gestell* once again to show itself. This is why the essence of technology must be thought of as a verb; it is not *what* technology is, but *how it is present*. Technology is present as *Gestell*, and thus conveys to human beings "that share in revealing which the coming-to-pass of revealing needs" (32). When humans think the essence of technology, they perforce also think about their own share in revealing. Thinking about the essence of technology thus allows them to address being, and brings to light the process of revealing once again.

According to Heidegger, "here and now and in little things . . . we may foster the saving power in its increase" (33). This happens through "essential reflection," which may prepare the way for the advent of a new destining. This essential reflection, however, cannot take the form of disengaging from technology in order that this will advent come to pass. "Because the essence of technology is nothing technological, essential reflection upon technology and decisive confrontation with it must happen in a realm that is, on the one hand, akin to the essence of technology and, on the other, fundamentally different from it" (35). This realm, according to Heidegger, is art. When art was conceived as *technè*, which for the ancient Greeks referred equally to both art and technology, it was seen as a bringing-forth that was not a challenging. An artwork is a human product that yet is not a standing-reserve, a human product that shelters its presencing into being explicitly within itself.

In "The Memorial Address" (written in 1959), which is effectively an addendum to "The Question Concerning Technology," Heidegger once again addresses the kind of reflection on technology that he sees as necessary. Therein, he links the reflection toward a particular attitude and behavior—a *comportment*—which is to be adopted to technical artifacts in order to make the saving power possible. He names this comportment "releasement" (*Gelassenheit*). According to Heidegger, human beings are becoming "encircled ever more tightly by the forces of technology. These forces, which everywhere and every minute claim, enchain, drag along, press and impose upon man under the form of some technical contrivance or other—these forces, since man has not made them, have moved long since beyond his will and have outgrown his capacity for decision" (Heidegger 1966, 51). They are beyond our power to control, according to Heidegger, because ultimately they depend on the way in which reality appears to us, namely, "as an object open to the attacks of calculative thought, attacks that nothing is believed able any longer to resist. Nature becomes a gigantic gasoline station, and energy source for modern technology and industry" (50). At the same time, Heidegger continues, technological devices have become indispensable to us, and it would be foolish and shortsighted to try to flee from them or to condemn them as the work of the devil.[5]

Human beings thus seem trapped. On the one hand technology impoverishes their relation to reality, while on the other hand the use of technology seems inescapable. This is precisely the dilemma that Heidegger wants to resolve through the notion of releasement or *Gelassenheit,* opening up the possibility of relying on them while not becoming enslaved to them. "We can use technical devices, and yet with proper use also keep ourselves so free of them, that we may let go of them any time. We can use technical devices as they ought to be used, and also let them alone as something which does not affect our inner and real core. We can affirm the unavoidable use of technical devices, and also deny them the right to dominate us, and so to warp, confuse, and lay waste our nature" (54). This simultaneous affirmation and denial of the technological world is a relation or comportment that Heidegger calls "releasement toward things" (*Die Gelassenheit*

5. This may be a direct reference to Jaspers's characterization of technology in terms of "demonism."

zu den Dingen). Releasement can be seen as "essential reflection" on the technological world in the literal sense of the word: it allows the essential meaning that is hidden in things to reflect in our thinking. "Having this comportment we no longer view things only in a technical way," and "we notice that while the production and use of machines demands of us another relation to things, it is not a meaningless relation" (54). The combination of letting technology be and holding oneself apart from it can create, Heidegger says, enough space to allow ourselves to stand in a realm in which the meaning of the technological world appears. We do not fully know, according to Heidegger, what this meaning is exactly, only that it is there. Because it is something that simultaneously shows itself and withdraws, Heidegger calls this comportment "openness to the mystery." Together with releasement toward things, openness to the mystery allows "the possibility of dwelling in the world in a totally different way" (55).

Heidegger works out this other way of dwelling in the world more concretely in "The Memorial Address" than in "The Question Concerning Technology." Whereas initially he only spoke about his hope for the arrival of a new way of revealing, which is actually *experienced* as a form of revealing and thus would be open to being, he now speaks about "a new rootedness in the soil," a new way of grounding oneself. In light of his affiliation with the Nazi party, from which he never distanced himself, this may well sound like the *Blut und Boden* nostalgia for the homeland in a new guise, but this impression would be mistaken. What matters for Heidegger is not the attachment of a people to its homeland, but rather the relationship that human beings have to their everyday environment. The technological way of disclosing the world, Heidegger means, leads human beings to lose their "home" in the world. He describes how human beings are "chained to radio and television" and are entranced by films that "give the illusion of a world that is no world." In summary, he writes, "All that with which modern techniques of communication stimulate, assail, and drive man—all that is already much closer to man today than his fields around his farmstead, closer than the sky over the earth, closer than the change from night to day, closer than the conventions and customs of his village, than the tradition of his native world" (48).

Releasement toward things and openness to the mystery must reverse this development and bring back human beings "to a new ground and foundation. In that ground the creativity which produces lasting works could strike new roots" (57). When human beings allow themselves to be swept along by technology, in short, they become alienated from their world—but a new comportment toward technology could allow the development of a new "home."

TO BE OR NOT TO BE—THAT IS THE QUESTION

Within the phenomenological orientation in classical philosophy of technology, Heidegger's work forms the counterpart of Jaspers's. Each of them represents one of the two directions—existential and hermeneutic—in which phenomenology has developed, and both have brought phenomenology in relation to technology. Just as Jaspers's existential approach invited critical reevaluation, so, too, does Heidegger's hermeneutical approach. Heidegger's views on technology, which in contrast to those of Jaspers are still influential, remain controversial. Some followers continue to use his language and arguments in denouncing the creeping uniformity of modern culture and the one-dimensionality of our thinking, while others attack the conservatism of his philosophy of technology and its contemporary partisans.

In this section I shall critically analyze Heidegger's philosophy of technology by analyzing in turn each of the three charges that are regularly lodged against it: that it is monolithic, abstract, and nostalgic. His work is said to be monolithic because he allows no room in his approach for an alternative technological practice; abstract because he single-mindedly focuses on technological thinking rather than on concrete technologies, and nostalgic because he often contrasts the present unfavorably with the exalted past. While I subscribe to the main points of these criticisms, they often remain too superficial in the ways they are given, because Heidegger's critics usually take an external point of view. Many forms of criticism of Heidegger, that is, are directed to the *conclusions* of his analysis rather than to the analysis *itself.* A serious critique of Heidegger's philosophy of technology cannot rest content with the claim that his picture of technology is monolithic, abstract, and nostalgic, but needs to show that this picture is *inadequate.*

Against the externalist perspective, I will take the alleged character of Heidegger's philosophy as a consequence of the way in which he thinks about technology. Therefore, I shall focus my analysis on the way in which Heidegger conceptualized technology and on the implications of this conceptualization. I will show that the charges of abstractness and of a monolithic character can be seen to arise from the fact that Heidegger, like Jaspers, employs a transcendentalist style of reasoning, which makes the foundations of his evaluation of technology problematic. In addition, I will make clear that Heidegger can support his nostalgic preference for a pretechnological relation with reality only by selectively navigating between two different approaches, one historical and one ahistorical.

Abstract and Monolithic: Heidegger's Transcendentalism

The American philosopher of technology Andrew Feenberg, working from the Frankfurt School tradition, is typical of those who critique Heidegger's abstractness and the monolithic character of his thought. He praises Heidegger for his refusal to see technology as a collection of neutral means and for his claim that, in some sense, we are not the masters of our own technology but in its grip. Nevertheless, he adopts an extremely critical position toward Heidegger. For Feenberg, Heidegger's philosophy of technology is too abstract, too remote from concrete technological practices. He writes, "The issue is not that machines are evil nor that they have taken over, but that in constantly choosing to use them over every other alternative, we make many other unwitting choices." He observes that Heidegger hopes for "a vaguely evoked spiritual renewal" but finds that his work is "too abstract to inform a new technical practice."[6] Heidegger's abstract reasoning means that "he literally cannot discriminate between electricity and atom bombs, agricultural techniques and the Holocaust" (Feenberg 1999, 187). This high level of abstraction makes Heidegger's philosophy of technology at the same time too general: by declaring modern technology a stage in the history of being, Heidegger leaves no room for the development of alternative technological practices. Technology simply has the essence of the *Gestell.* Not technology itself, but our way of thinking and disclosing reality will change in the next epoch of the history of being—one for which we can only wait expectantly.

While I share this critique, I think it requires more support than Feenberg provides. The simple claim that Heidegger is too abstract and monolithic to help inspire a new technological practice is in itself insufficient to be an adequate critique. Feenberg ignores Heidegger's claim that the essence of technology is itself nothing technological—and when Feenberg writes enthusiastically about the need to develop a new technological practice, any Heideggerian will be able to throw that remark back at him, pointing out that this would be just another expression of the will to power and would simply reinscribe the *Gestell* in a more sophisticated way. Heidegger does not want an alternative technological practice, for he does not want to think on the ontic level (that of beings), but rather on the ontological level (that of being). Were Feenberg to engage in a serious discussion with Heidegger, *either* he would have to claim that *only* an ontic approach to technology is adequate, *or* he would have to critique Heidegger's ontological analysis. A

6. See paragraph 13 of Feenberg (2004).

true critique cannot be based on the consequences of an approach, but must rather be directed at the approach itself.

Implicitly, however, Feenberg does make a remark that points the way toward a more fundamental critique. The reason, according to Feenberg, that leads Heidegger to lump together agricultural techniques and atomic bombs is that he sees technology "merely as different expressions of the identical enframing" (187). Here Feenberg is addressing the way in which Heidegger looks at technology—not at the specific technologies of our world, but rather at technology as a form of world-disclosure—and touches on a serious problem in Heidegger's analysis. Heidegger's answer to the hermeneutical question of the way in which technology discloses the world to human beings is too abstract because it loses contact with technology itself. Heidegger does not approach technology (ontically) in terms of concrete technological artifacts, but (ontologically) as a form of world-disclosure. And the manner in which technologies disclose reality, according to Heidegger, is not determined by these technologies themselves, but rather by "being." The "revealing" of technology always already receives its form from the ruling way of the "unconcealment" of the *Gestell.*

Technology thus does not itself *create,* according to Heidegger, a specific form of world-disclosure, but is instead a *manifestation* of one. Even before specific technologies mediate our relation with reality and disclose it to us, it has already been determined in what way reality can show itself. For Heidegger, it is not technologies that disclose reality but the ruling way of unconcealment. Technologies are only manifestations or expressions of a form of disclosing reality, instead of its source. Feenberg's reference to Heidegger's abstractness thus needs to be understood as a critique of Heidegger's ontologizing of technology; the consequence of looking at technology only ontologically is that concrete, ontic technologies drop out of sight.

But is this critique right? Heidegger, after all, provides examples that always involve specific technologies. Consider his analysis of the hydroelectric plant:

> The hydroelectric plant is set into the current of the Rhine. It sets the Rhine to supplying its hydraulic pressure, which then sets the turbines turning. This turning sets those machines in motion whose thrust sets going the electric current for which the long-distance power station and its network of cables are set up to dispatch electricity. In the context of these interlocking consequences of the ordering of electrical energy, even the Rhine itself appears as something at our command. The hydroelectric plant is not built into the Rhine River as was the old wooden bridge

> that joined bank with bank for hundreds of years. Rather the river is dammed up into the power plant. What the river is now, namely, a water power supplier, derives from out of the essence of the power station. . . . But, it will be replied, the Rhine is still a river in the landscape, is it not? Perhaps. But how? In no other way than as an object on call for inspection by a tour group ordered there by the vacation industry. (Heidegger 1977a, 16; translation slightly modified in order to give a literal translation of the original text)

The way Heidegger elaborates his example here appears to be the result of his analysis of the way in which a specific hydroelectric plant discloses the environment in which it is placed. His remarks about the technological unlocking of nature appear to be based on an empirical inquiry into an actual hydroelectric plant that, because it was built directly into the Rhine, makes the Rhine nothing more than a means for electrical generation. Heidegger appears to be proceeding not ontologically but ontically.

Looks deceive, however. Heidegger not only claims that "in the context of" this technology the Rhine appears as a water power supplier, but refers to the plant and its associated technologies as "the interlocking *consequences* of the ordering of electrical energy." His words reveal that, for him, what is happening is not that the construction of an electrical generating plant has brought about the transformation of the Rhine into standing-reserve, but rather the other way around—that the unlocking of the Rhine as standing-reserve has brought about the construction of an electrical power plant in it. This is underscored by his remark about the tourist industry, which seems to form part and parcel with the hydroelectric plant. The Rhine *cannot* be disclosed otherwise than as standing-reserve: with the *Gestell* as the reigning way of unconcealment there is only one manner in which things can be revealed. That the Rhine has the character of standing-reserve is thus not due to the actual technologies that disclose it as such. The relation between the hydroelectric plant and the Rhine is precisely the opposite: only because the river, like the rest of reality, shows itself as standing-reserve can something like a hydroelectric plant be designed and built.

Is this an exceptional case, or is it representative of Heidegger's approach? According to Günther Seubold, who has made a fastidious analysis of the role of technology in Heidegger's oeuvre, it is the latter. Seubold asserts—without regarding this as something negative—that a conditional relation (*Bedingungsverhältnis*) exists between technological revealing and specific technologies. This relation appears not only in "The Question Concerning Technology" but also in texts such as *What is Called Thinking*, where Heidegger writes, "Our age is not technological because it is the age of the machine; it is rather the age of the machine because it is the technological age."[7] Seubold therefore concludes that technological devices "can only be made when the technological way of revealing has already set in" (Seubold 1986, 195).

Seubold's further working out of the relation in Heidegger's philosophy of technology between the ontic and the ontological levels illustrates this still better. "To what extent is the invention of a machine due to the essence of technology?" Seubold asks. He answers that the essence of modern technology has the characteristic that "to an increasing extent it calculates everything and searches a cause and ground for everything; that it forces nature to show itself in terms of causes and mechanisms." And this is precisely what makes possible the invention of machines: only when reality is disclosed in this way can there be the kind of interventions into reality that make it possible to design and build machines. As Seubold writes, "the machine could never have been invented unless nature were already disclosed as a web of causes and effects" (196).

But the connection between "ontic technology" and "ontological technology," according to Seubold, is more than that the latter is the condition of the former; there is also an internal connection between the two. The ontic technology with which we are familiar is not an arbitrary collection of artifacts; rather, (ontological) technology "shapes the specific means which are characteristic of it." Technological artifacts, as it were, according to Seubold, reflect the *Gestell:*

> The technological power which human beings have over things is manifest in means, instruments, and machines. These refer to the technological comportment which human beings have to things, and are not related to this comportment externally or indifferently. The Heideggerian interpretation of the essence of technology as a way of revealing does not at all lose sight of technological means. It does indeed pay attention to them, but maintains that these devices are ultimately not fundamental, but instead refer to another ground on which they depend, namely the ontological happening of "forcing into appearance" and "setting-upon." Only means of this type refer to this way of revealing; other means and ends depend on other ways. (206)

7. See Seubold (1986, 195). A similar remark can be found in Heidegger's "The Age of the World Picture": "Machine technology remains up to now the most visible outgrowth of the essence of modern technology, which is identical with the essence of modern metaphysics" (Heidegger 1977c, 116).

The example of the hydroelectric plant is thus not idiosyncratic, but instead typifies Heidegger's approach to technology. Specific technologies are secondary, for him, with respect to the "essence" of technology, and only from this "essence" can technology be understood. Heidegger does not look at technology ontically but ontologically, regarding the former as derived from the latter. When he speaks about technology, he means not specific technologies but rather the *Gestell.* This makes his work of necessity "abstract," in the sense that he abstracts from specific technological practice.

In this way, Feenberg's otherwise external criticism of Heidegger's abstractness is brought in closer relation to Heidegger's thinking. But this does not yet make a genuinely adequate critique. Feenberg is correct to claim that Heidegger's work is too abstract to inspire a new technological practice—but if Heidegger is right in claiming that each technological practice depends on the *Gestell* as a specific "sending of being," there *cannot* be an alternative technological practice. Either one has to accept Heidegger's ontological account and agree that it is not necessary to look at concrete technologies, or one needs to give arguments for the position that, in so doing, Heidegger overlooks important aspects of technology. As should be clear by now, I am sympathetic with the latter approach, and shall show how it can be clarified, as in the case of Jaspers, by examining the transcendentalist style of Heidegger's thinking.[8] Because Heidegger looks at technology in an ontological way—that is, because he sees technology as an expression of a sending of being—he reduces concrete technologies to the sending of being that forms their condition of possibility. The sending of being as *Gestell* thus *precedes* the specific technologies that surround us. Heidegger then absolutizes these conditions of possibility; he speaks about them as if he were speaking about the technologies themselves, and even *names* this sending of being "Technology."

Heidegger is thus "abstract" in a quite specific manner: he abstracts from concrete technology by reducing it to its condition of possibility. Accordingly, he only sees those aspects of technology that follow from those conditions. When specific technologies are seen as nothing more than the material realization of the ruling way in which reality is disclosed, depending on a prior conception of being, then these technologies only seem to allow this specific way of disclosing reality. Moreover, because Heidegger sees the underlying meaning of being as beyond the reach of humans to alter, his conception of technology at the same

8. As I remarked in the Introduction, the term "transcendentalism" has been used before to characterize Heidegger's philosophy; see note 3 to the Introduction.

time has a monolithic character. Human beings can only await and *prepare* for the arrival of a new sending of being, but cannot *bring it about.*

But it is idle merely to criticize Heidegger for being abstract and overly general; the question we should ask is what more there is to say about the relation between technology and meaning. Should we follow Heidegger in his claim that the *Gestell* is the only form of "unconcealment" in our world? The answer, I think, is no; furthermore, I think that another approach to technology is possible that does more justice to the actual role of technology in our culture and our everyday lives. While Heidegger may be right that a specific, technological way of interpreting reality (on the ontological level) is required for modern technology to come about, we should also conclude that the role of technology (on the ontic level) in our culture cannot be understood in terms of this specific way of interpreting only. When they are used, technologies may make it possible for human beings to have a relation with reality that is much richer than those they have with a manipulable stock of raw materials.

Modern communication technologies, for instance, make it possible for human beings to interact with others outside of their immediate environments without approaching these people from a "will to power." And medical technologies, to be sure, depend on an interpretation of the human body in which it is a manipulable material, but this interpretation in most cases serves another purpose: care and respect. I am aware that these counterexamples are neither shocking nor surprising, but precisely their triviality makes clear that there is a blind spot in Heidegger's approach. The price of Heidegger's a priori construction that there is always one way of unconcealment that holds sway, which, in our epoch, has the character of the *Gestell,* is a highly impoverished image of technology. A telephone does not reveal one's interlocutor as standing-reserve but as a unique individual person, while a medical instrument such as an MRI can reveal a patient not only as a body permeated by causally linked connections, but also as someone whose life is more than controllable but also intrinsically valuable.

The Belgian philosopher of technology Gilbert Hottois has argued that technology is characterized not only by the fact that it depends on an *interpretation* of reality, but also by the fact that it *intervenes* in reality—meaning that it is not adequately understood if it is grasped only in terms of interpretation because this reduces it to the domain that it actually transcends (Hottois 1996, 15–17). Technology "does something," it has effects, it works—and these workings cannot be reduced to interpretation. For an adequate analysis of communications technologies—to return to this example once again—it is not enough to say that these technologies were made possible by a particular way of disclosing reality. We should also analyze in what ways they, in their functioning, create new ways of access to reality for human beings. Heidegger, as it were, annexes technology to philosophy, reducing it to a form of thinking and then analyzing that thinking as if it were a brand of philosophy.

The philosophy of technology needs to take specific technologies more seriously than Heidegger does; in Hottois's terms, it needs to try to elucidate its *operativity.* A philosophical analysis of the role of technology in the modern world cannot rest with *reducing* technology to forms of interpretation, but needs to devote its attention as well to the ways in which specific technologies and artifacts help to *shape* specific forms of praxis and interpretation. It needs to think "forward" rather than "backward" about technology. Only then will the hermeneutical perspective genuinely be able to make visible the role that technology plays in the way in which "reality" appears to us. Technology will then not appear as a form of alienation from the "full richness" of reality, but as a form of mediation of the relation between human beings and reality. A philosophy of technology that aims to answer the hermeneutic question asking what role technology plays in human interpretations of reality should analyze this very mediating role of technology, in close contact with the empirical reality of technology.

In this way, the abstract and monolithic character of Heidegger's philosophy of technology will be overcome, for this approach will take seriously specific technologies rather than seeing them as simply expressions of an a priori sending of being. Against Heidegger's conviction, as set out in "The Question Concerning Technology," we *do* have to look amid the "rods, pistons, and chassis" to understand technology in an adequate way (Heidegger 1977a, 20). And against Heidegger's claim that the essence of technology is "nothing technological," we should develop an analysis of technology that does take concrete technological tools, instruments, and devices seriously.

Nostalgia: Heidegger's Two Approaches

The second avenue of criticism that I want to develop concerns the references to Heidegger's philosophy of technology as nostalgic. In his works, Heidegger creates a sharp opposition between modern technologies and traditional techniques. The former he judges negatively, thanks to the one-dimensional and dominating way of revealing with which they are linked, while the latter he assesses to have "released" and "let things be." Nostalgia, surely, is not necessarily bad—nothing is wrong with it per se as long as one can give good reasons for it. But Heidegger's arguments for valuing traditional techniques over modern ones appear to be susceptible to fundamental criticism.

One critic who charges Heidegger with nostalgia is Don Ihde (1993b, 103–15). According to Ihde, Heidegger's descriptions of traditional and modern technologies are colored by the "romantic thesis" that traditional technologies are preferable to modern ones (106–7). But when one looks carefully at Heidegger's examples, the traditional technologies he champions turn out to exhibit a dimension of domination and control as well as the modern ones, while the modern technologies he derides also exhibit a degree of "letting things be."

Ihde provides two examples of the selective character of Heidegger's descriptions of traditional technologies. One involves Heidegger's comparison between the windmill and the hydroelectric plant in "The Question Concerning Technology," while the other concerns his example of the Greek temple in "The Origin of the Work of Art." The hydroelectric plant, according to Heidegger, is connected with the disclosure of the Rhine as standing-reserve—the Rhine is "ordered" to supply energy. The windmill, by contrast, does not disclose the wind in the same way: "Its sails do indeed turn in the wind; they are left entirely to the wind's blowing. But the windmill does not unlock energy from the air currents in order to store it" (Heidegger 1977a, 14). Ihde asks himself what Heidegger would say about another traditional technology, the waterwheel, and concludes: "While this [the hydroelectric plant's] production of energy does contrast with the old windmill—which can turn only when the wind blows and which thus seemingly lets the wind 'be'—it does not, in principle, differ from the smaller dam on the stream which allows the waterwheel in turn to grind the peasant's wheat. To allow this example as a 'good' technology does not, to my mind, prevent seeing nature as resource well except in its lack of a larger interconnectedness with the electrical grid" (Ihde 1993b, 107). For Ihde, there is no difference in kind, only difference in degree, between a hydroelectric plant and a waterwheel; even the waterwheel "sets" a river "to supplying its hydraulic pressure" (Heidegger 1977a, 16).

Ihde could also have illustrated his claim that Heidegger's description of traditional technologies is biased by discussing another artifact Heidegger contrasts with the hydroelectric plant, namely "the old wooden bridge that joined bank with bank for hundreds of years" (16). Such a bridge, Heidegger claims, is "built into" the Rhine, while the Rhine is "dammed up into" the hydroelectric plant. By the phrase "built into the Rhine River," Heidegger evidently means something to the effect that it "respects" the Rhine in some way, or "does not force the Rhine to reveal itself in a way that is not proper to it." But can't this also be said of the hydroelectric plant? After all, the Rhine allows the power plant to be inserted in it, and the plant would not function if the Rhine did not cooperate. The power plant, therefore, is "built into" the Rhine as well, since it must be structured in accordance with the Rhine's currents in order to work. What, then, is essential about the Rhine that the bridge respects and the power plant does not? If it is that the Rhine is left to flow undisturbed, then clearly the power plant indeed dams it up. But if what is essential about the Rhine is that it separates two shores, then the bridge has been doing violence to it for centuries. Viewed from this perspective, doesn't the bridge represent a challenging way of dealing with the Rhine, in which its currents are not thankfully treated as a gift but rather as an obstacle to getting from one side to the other?

Ihde's second example of the romantic coloring of Heidegger's preference for traditional technologies is his "phenomenology of the Greek temple" in "The Origin of the Work of Art." In that essay, Heidegger describes how a temple discloses a landscape: it discloses the rocky ground on which it stands through "that rock's clumsy yet spontaneous support," the storm that rages about it, the light that gleams off of it and "makes visible the invisible space of air." The temple does not "set in order" the landscape, but rather lets it "come to be." According to Ihde, however, Heidegger's description of this temple could be exchanged for a completely different one. Historian J. Donald Hughes, for instance, has pointed out that the mountain peaks of Attica behind the Acropolis are so bald and dry because the Greeks deforested them of trees in order to build the Parthenon. Why does Heidegger leave this relation between the temple and the landscape out of his description? Why doesn't he point out that the temple at the same time discloses the surrounding woods as raw materials with which temples can be built? Heidegger conceals here what Langdon Winner calls the "politics of artifacts," the social and political consequences of the artifact when it is manufactured and used (Ihde 1993b, 111). According to Ihde, then, concealing is thus not only an activity of "being," but also of Heidegger himself (104–5).

According to Ihde, Heidegger is selective in his descriptions not only of traditional technologies, but of modern technologies as well. Ihde supports this claim by critiquing Heidegger's famous description of the bridge in "Building, Dwelling, Thinking." In his late period, during which he wrote this essay, Heidegger creates an ever starker contrast between two ways of disclosing reality: as "standing-reserve" and as "fourfold" (*Geviert*). The fourfold, for Heidegger, is the world as it is gathered by "earth and sky, divinities and mortals" (Heidegger 1971b, 153).[9] These four components form the dimensions that open up the realm in which human beings experience their world. They are made visible by *things,* which refer to the earth from which they are made, the sky under which they rest, the mortals who concern themselves with them, and the gods who can be thanked for them. A bridge, for Heidegger, is a good example of such a "thing":

> The bridge swings over the stream "with ease and power." It does not just connect banks that are already there. The bands emerge as banks only as the bridge crosses the stream. The bridge designedly causes them to lie across from each other. . . . With the banks, the bridge brings to the stream the one and the other expanse of the landscape lying behind them. . . . The bridge *gathers* the earth as landscape around the stream. Thus it guides and attends the stream through the meadows. . . . Even where the bridge covers the stream, it holds its flow up to the sky by taking it for a moment under the vaulted gateway and then setting it free once more. The bridge lets the stream run its course and at the same time grants their way to mortals so that they may come and go from shore to shore. . . . The bridge *gathers,* as a passage that crosses, before the divinities—whether we explicitly think of, and visibly *give thanks for,* their presence, as in the figure of the saint of the bridge, or whether that divine presence is obstructed or even pushed wholly aside. (152–53)

Heidegger's description of the way in which the bridge reveals the river and the landscape contrasts sharply with the way in which a piece of modern technology reveals; recall the description of the hydroelectric plant from "The Question Concerning Technology" that only revealed the Rhine as a water power supplier.[10]

9. See the fourth section of this chapter, titled "Heidegger and Things," for a further explication of the concept of the "fourfold."

10. I follow here Ihde's argumentation in *Postphenomenology* only partly, since it contains a mistake. In "Building, Dwelling, Thinking," that is, a modern bridge also plays a role—a bridge over a highway—and Ihde erroneously reads in Heidegger's description of it a pejorative view of modern technology. Heidegger writes, "The highway bridge is tied into the network of long-distance traffic, paced as calculated for maximum yield" and facilitates "the lingering and hastening ways of men to and fro," but it is still related to the fourfold "whether that divine presence is obstructed or even pushed wholly aside" (Heidegger 1971b, 152–53). From this, Ihde concludes that, for Heidegger, the modern-technological bridge does not reveal the fourfold, while the old bridge does (Ihde 1993b, 109). This conclusion, however, conflicts with Heidegger's claim, in "Building, Dwelling, Thinking," that all the bridges he describes reveal the fourfold, including even the modern bridge. Heidegger claims only that, for modern humans, the presence of the fourfold is ignored. The hydroelectric plant

For Heidegger, only pretechnological or traditional artifacts such as bridges, jugs, and chalices reveal reality in terms of the fourfold. Ihde finds this too limited; why could not modern technologies also reveal the world in terms of the fourfold? Consider, for instance, Ihde's provocative "phenomenology of the Shoreham nuclear power plant," which was constructed but never operated on the shores of Long Island Sound:

> Seen while sailing in Long Island Sound, on the horizon stands the stark super-silo, light green topped, of the Shoreham nuclear plant. Standing there, it brings to presence the very contrast between the seemingly featureless sandhill earth with the sky. It stands at and defines the contrast, too, between the sea and the shore, which without its focal presence would also be featureless lines along the horizon. . . . In its gathering, the nuclear plant makes the fishy life of the Sound to appear—as drawn to the warmer exhaust waters of the plant, but to be placed in danger of a leak. . . . It channels the community into its pathways. . . . It reveals the hastening which would be needed to evacuate its wastes.[11] (Ihde 1993b, 111–12)

In a word, Ihde's critique boils down to the fact that Heidegger's nostalgia rests more on romance than on argumentation. Heidegger makes plausible neither why traditional technologies could not reveal reality as raw materials nor why modern technologies would not be able to reveal the fourfold.

To Ihde's critique I would add that Heidegger's nostalgic conception can be understood as a product of his transcendentalist approach. Because Heidegger reduces technology to the way of disclosing reality that underlies it, he discerns only that specific way of disclosing when he analyzes technology; he finds only what he has already projected. Because he conceives technology as a manifestation of a historically specified form of unconcealedness, he must project earlier forms of unconcealedness onto traditional technologies (in the ontic sense) and can only see the unconcealedness as *Gestell* in modern technologies. In earlier sendings, being was not yet concealed, and because of this, Heidegger also lets the specific technologies of these periods help to reveal being. In the contemporary

---

of "The Question Concerning Technology" is a better example to contrast with the old wooden bridge. In these two examples, one is clearly tied to disclosure as standing-reserve and the other one to disclosure as fourfold.

11. Ihde has left out only the divinities, who would be thanked for the plant and for saving mortals from the energy crisis.

sending of being, however, being hides itself, which Heidegger connects with the way in which modern technologies reveal reality. Heidegger, that is, does not investigate the differences between the ways in which traditional and modern technologies reveal reality, but instead infers the way of revealing from the sending of being that underlies the specific technologies.

Yet a more fundamental critique can be lodged against Heidegger's nostalgia. For Heidegger supports his nostalgic preference for traditional technologies not only through selective descriptions but also through selectively employing two different approaches in his analyses. When analyzing traditional artifacts he uses an ahistorical perspective, while he approaches modern technologies using a historical perspective. This becomes clear when we look more closely, once again, at his analysis of the contrast between the hydroelectric plant and the old waterwheel in the Rhine. On the one hand, in the former case Heidegger conceives the revealing connected with technology as a contingent stage in the history of being: the hydroelectric plant reveals reality as standing-reserve because the ruling way of unconcealedness is that of the *Gestell.* On the other hand, he conceives the revealing connected with technology as an intrinsic characteristic of objects: as a thing, the bridge reveals the world in terms of fourfold, and it does so in every epoch, however much this may be "obstructed or even pushed wholly aside" in specific epochs. The way in which a technological object reveals reality is, therefore, in the first instance *historically* sent by being, while in the second instance it is represented as a *fundamental* event that can be veiled by a purely technological way of thinking.

Only by making use of these two perspectives simultaneously can Heidegger support his nostalgic preference for traditional technology. And in light of the radically historical character of his work, his preference for traditional technology is remarkable indeed. One can be nostalgic only when one thinks that something *essential* has been lost, and that becomes problematic precisely when one thinks *historically,* for then something can only be essential within a historical context rather than ahistorically. From a purely historical perspective, classical *technè* and modern technology would be historical phases in the relation of humans to being, and neither could claim to be more fundamental than the other. Both the mill and the hydroelectric plant would then equally represent a contingent manifestation of "coming into being." Inversely, when thinking from a purely ahistorical perspective, there is no reason why the fourfold would not be gathered by modern technologies such as the hydroelectric plant as well. The claim that traditional technologies reveal the "coming into being" of entities and that modern technologies do not presupposes both a historical and an ahistorical understanding of being at the same time. The claim treats "being" not only as an always historical fleshing out of what it means "to be," but at the same time as a fundamental and ahistorical happening of "coming into being" that either can be experienced (in the age of *technè*) or remain unnoticed (in the age of modern technology).

In Heidegger's philosophy of technology, therefore, being is defined on the one hand as the interpretation of "what it means 'to be,'" which changes over time and which is always present at the background of human relations with the world; and on the other hand as the "happening of coming into being," the event of revealing that occurs continuously but that is not always noticed and thought. Being has both intrinsic and historical characteristics. Intrinsically, and therefore ahistorically, it is an "event," but the way in which this event shows itself is historical. *That* reality comes into being is unchangeable, but *how* this event shows itself is contingent. Heidegger, after all does not lament a *loss* of being but rather its *forgetting*. Being is always at work whenever and wherever beings manifest themselves, but in our present age the character of their manifestation does not allow that their "being" is noticed.

An ambiguity thus eventually emerges in Heidegger's conception of being.[12] His revitalizing force in philosophy is due precisely to his historical approach; he opened the door to postmodernism by approaching being as changeable rather than static, and thus the "essence" of things as contingent, resting on a historically determined conception of being. Heidegger forces us to relinquish the illusion that we can penetrate to "essences"; which is, as Eric Bolle puts it, a "farewell to what never was" (Bolle 1985, 116–19). But this "postmodernism *avant la lettre*" is also bound up with a romantic motif, for Heidegger laments the fact that being, in the form it has taken since Plato, has concealed itself. And "being" here

12. When Michael Zimmerman, in his book *Heidegger's Confrontation with Modernity*, discusses Charles Guignon's charge that Heidegger is guilty of "foundational historicism," he points to a similar kind of ambiguity in Heidegger's philosophy (Zimmerman 1990, 147–48). Guignon points out, that is, a tension between "foundational" and "historical" thinking in Heidegger's approach to human existence (*Dasein*). By showing that each epoch has its own framework that gives meaning to the relation between human existence and beings, Heidegger on the one hand points to a historical, transcendental structure to human existence (the understanding of being precedes both human existence and its relation with beings), while on the other hand he historicizes that existence (human existence itself receives its meaning through the historically specified understanding of being, and human beings have no direct access to their own being). This tension, according to Guignon, is what brought Heidegger to the *Kehre:* his turn from an analysis of the structures of *Dasein* of *Being and Time* to the play of being of his later work, in which no transcendental analysis of human existence is necessary in order to understand how being announces itself. The ambiguity in Heidegger's philosophy to which I am referring takes place in the period following the *Kehre*, in which the same problem appears to recur, but with being rather than existence conjoining the ahistorical and historical elements.

is not a philosophical construction to explain that what counts as reality for human beings depends upon our interpretation of what it means "to be." It rather is an actual happening, an experienceable event (*Ereignis*): coming to be, emerging out of concealment into unconcealment.

Within the framework of Heidegger's own conception of being this ambiguity does not pose a problem. Its two poles do not necessarily conflict: being is intrinsically (or ahistorically) a happening or event that can only show itself in (historically) contingent ways. The problem arises only when Heidegger applies his ambiguous conception of being in the context of his philosophy of technology, by connecting the two aspects of being *selectively* with modern and traditional technology. His ambiguity then turns into the application of a double standard.

What is decisive when he reflects about modern technology is the historical dimension of being—*how* "coming into being" is understood in different epochs—while when it comes to reflecting about traditional technology the ahistorical dimension is decisive—*that* reality "comes into being." The hydroelectric plant figures as an exponent or outcome of a specific "sending" of being; it could only be built because reality shows itself as standing-reserve, and the "event" of coming into being is forgotten. By contrast, the mill is an exponent of being itself; it is a thing that, free of any and all historical contexts, makes visible that being is an event—even though in some epochs this visibility remains unnoticed. Without the selectivity in Heidegger's approach either the hydroelectric plant would have to reveal the fourfold as well or the bridge could not be understood as a thing that essentially (ahistorically) reveals the event of coming into being. In other words, Heidegger's understanding of modern technology in terms of alienation, in which something originary gets lost, would not have been possible without applying a double standard.

Heidegger develops his two standards in two different contexts. The historical standard, which forms the background from which he analyzes modern technology in "The Question Concerning Technology," approaches technology as "producing" or "making." The ahistorical standard, from which he analyzes traditional technologies in "Building, Dwelling, Thinking" and "The Thing," is addressed to the *objects* that are technically produced. Understanding technology in terms of production allows Heidegger to understand modern technology as a stage in the history of being. In production, after all, objects come into being, which makes it possible to treat these objects as expressions of the ruling way of revealing. By contrast, Heidegger only connects his ahistorical conception of being with an understanding of technology as technically produced objects. This allows him to create a romanticized point of reference in the past. The products of traditional crafts stand in an ahistorical relation to being; they bring about "coming into being" itself by gathering the fourfold.

It might be objected that Heidegger nevertheless did approach traditional technology in a historical way as well. The windmill in "The Question Concerning Technology," for instance, did not "order" wind energy yet because it remained dependent on the wind's blowing. The mill can be understood as the outcome of another meaning of "being"; it was not built on the basis of an understanding of nature as fully available for human manipulation. Yet the confusion between two different approaches recurs once again here, albeit in a different guise. The mill might indeed figure historically as an exponent of another phase in the history of being—but what it makes visible is precisely the *ahistorical* aspect of being, namely, that being is essentially an event of "coming into being." The mill makes this visible because its functioning is beyond human control. Therefore, the mill reveals not so much the specific way in which being is understood in a specific historical context, but rather the event of "coming into being" itself. The mill is not a *consequence* of a historical epoch in the history of being, as is the hydroelectric plant according to Heidegger; rather, it makes visible being *itself*.

Ihde's charge that Heidegger's nostalgic contrast between tradition and modernity is untenable can now be more fundamentally supported. Heidegger measures tradition and modernity with different scales. When he speaks about traditional technology, he does so in ahistorical terms; when he speaks about modern technologies, he is a historian. Heidegger's "path of thinking," as Pöggeler indicates, i.e., Heidegger's philosophy (Pöggeler 1991), appears to be a two-way street.

To Be or Not To Be

Heidegger's hermeneutic philosophy of technology appears to raise two problems. His approach is abstract and monolithic in that he reduces specific technological artifacts to historical products of the history of being, a history that develops itself beyond human control; and he has a nostalgic preference for traditional technology, supported by an ambiguous approach to the relation between technology and being. When he compares specific technologies of past and present with each other, he applies two different standards, reserving a historical perspective for an analysis of modern technologies and an ahistorical perspective for traditional technologies.

Against Heidegger's ontologizing of technology, it became clear that the hermeneutical perspective could be carried out otherwise as well, in a way that does not proceed backward to an investigation of the conditions of possibility on which modern technology rests but forward to an inquiry into how specific technologies disclose reality. Philosophy of technology needs to get closer to the technologies themselves, so to speak. On an intuitive level it became clear that technology does not necessarily only imply a reduced relation to reality; it also makes possible new relations to it. Elaborating this intuition will require an approach to technology in terms of concrete technological artifacts rather than a reflection of the "ruling interpretation of the meaning of being" that lies behind it.

Paradoxically enough, Heidegger's philosophy itself offers points of departure for such an approach of technology in terms of artifacts. If ever there was a philosopher who could not be accused of thinking too little about "things," it is surely Heidegger. Instead of immediately going elsewhere, then, I shall first analyze his texts about "things," which run throughout his career, though with decreasing relevance to the kind of philosophy of the technological artifact that I mean to develop. His early writings especially is a fruitful point of departure for a philosophy of technology that takes artifacts seriously, both as a material culture in which reality acquires new meanings and as objects that provide human beings with new means of actualizing their existence.

HEIDEGGER AND THINGS

My discussion of the classical philosophy of technology culminated in the claim that the philosophies of technology of Jaspers and Heidegger reveal the necessity of a "turn toward things." In the following section I shall attempt to articulate the first steps of such a turn through pointing out what contributions Heidegger's work can make toward this effort. The most promising points of departure are found in his early work.

The views of Heidegger I have discussed so far are often referred to as if they represent his entire philosophy of technology, but they should properly be regarded as his second philosophy of technology. For in his early principal work, *Being and Time*, and other early works we can find another, albeit implicit, philosophy of technology. This earlier view offers, in particular, an extensive analysis of the role of equipment in the relation between human beings and their world, which contrasts sharply with his later analysis. For, according to the earlier Heidegger, instead of *reducing* the relation between human and world, technological artifacts generate specific forms of access to the world for human beings. This analysis offers fruitful points of departure for the formulation of a philosophy that seeks to understand technology through its artifacts.

The stark contrast between Heidegger's early and late approaches to technology is linked with the so-called *Kehre* or turn that his philosophy took. "The Question Concerning Technology" and "The Memorial Address" were composed after the *Kehre; Being and Time* before. The turn involved a change in his manner of approach to the question of being, the question around which his philosophy revolved from first to last. Initially he attempted to think being via an analysis of Dasein, or human being; because humans were concerned about their own being and had some understanding of it, Heidegger sought access to being itself via clarification of the human way of being. Later, however, he tried to think being itself, in terms of the ways in which it could be understood and in which it manifests itself.

In this section I shall analyze the development of Heidegger's thinking about things, including technological artifacts, in connection with this turn in his thinking.[13] In order to do this, I shall consider six texts, following the order in which they were written. First I shall treat Heidegger's analysis of tools and equipment from *Being and Time* (1927). Then I shall take up the development of his "thinking about things" from "The Origin of the Work of Art" (1935), which can be regarded as a transitional work en route to "The Thing" (1950) and "Building, Dwelling, Thinking" (1951), and finally to "The Question Concerning Technology" (1953) and "The Memorial Address" (1955). This analysis will make clear that Heidegger's *Kehre* can be held responsible for bringing about the transcendentalist perspective in his thinking about technology. I shall show how Heidegger reduces the thing more and more to "being," which only needs the thing as a medium to manifest itself. By looking at the transformation in Heidegger's thinking from the perspective of his approach to things, it becomes possible on the one hand to show the importance and the specific character of a "philosophy of technology from things," and on the other hand to take a first step toward finding a point of departure for such an approach of technology.

Tools as Providing Access to the World

In *Being and Time,* Heidegger works out his own, much extended version of Husserl's phenomenology. While Husserl's phenomenology consisted of the transcendental elucidation of the relation between consciousness and reality, Heidegger concerned himself with the relation between human existence and its world. He tried to strip phenomenology of its tendency to become a philosophy

13. I follow the chronology of Richardson (1963, 678–80).

of consciousness, and sought to reconstruct it as an analysis of the way in which human beings act in and experience their dealings with their world. Heidegger describes the relation between humans and world not as "intentionality," as Husserl had, but as "being-in-the-world."[14]

In Heidegger's analysis of "being-in-the-world" things play an important role; indeed, in the form of tools they make it possible that relations between humans and the world come about.[15] But to bring this role of tools to light requires a particular kind of description, one that does not approach them from the outside as objects to be described or analyzed, but from within in terms of the kind of presence they have in our everyday dealings with them. For the way in which things are present to human beings in everyday life precedes the way we describe or analyze them. Explicit attempts to name or describe things pass over this everyday presence, in which human beings take up and use these things as if for granted, without explicitly remarking upon them as things. Someone who hammers a nail into the wall is not focusing his or her attention on the hammer as such, but is rather absorbed in a practice within which hammer, nail, and wall each play natural roles. The ancient Greeks called things *pragmata,* this name pointing precisely in the direction of the approach Heidegger wants to take. Things belong to the realm of praxis and must be approached pragmatically; i.e., in relation to the behavior of human beings. Such behavior, as a way of being-in-the-world, Heidegger calls *besorgen:* heeding or taking care, the concernful dealing with the world that gives form to human existence. The objects that play a role in such heeding are called *Zeug:* tools, equipment, useful things.

Heidegger sets out to discover the way of being of these useful things. What makes a tool a tool? From the perspective of everyday life, it would be a mistake to try to find the answer with the aid of the classical concepts with which philosophy has traditionally tried to understand things, such as substance, materiality, and extension. This would approach things from an abstract and analytical relation to them, rather than from the everyday dealings that precede such a relation. From the perspective of praxis, a useful thing is "something in order to . . . "; it is useful, helpful, serviceable (Heidegger 1996, 64). Just as consciousness, according to Husserl, is always "consciousness of . . . ," so tools and equipment never exist simply in themselves, but always refer to that which can be done with them. What makes a tool or piece of equipment what it is, is that it makes possible a practice. But a remarkable feature of the way tools are present is that they withdraw

14. See Ihde (1979, 104, 116–17). Ihde speaks even of "existential intentionality."
15. Heidegger carries out this analysis in paragraphs 15 and 16 of *Being and Time.*

from, or hide in, as it were, the relation between human beings and their world. Generally, human beings do not focus on the tool or piece of equipment they are using, but on the work in which they are engaged. A person who writes is not directing his or her attention to the pen or keyboard, but to the text. The more attention that a tool or piece of equipment requires, the more difficult it is to do something with it. The way of being a tool or piece of equipment has when in use Heidegger calls *Zuhandenheit:* "handiness" in the newer translation, or "readiness-to-hand" in the older.

But even as tools and pieces of equipment hide themselves in the relation between human beings and world, they also shape it. The "in order to" of a tool or piece of equipment calls for a particular practice, which discloses the world in a particular way. First of all, according to Heidegger, tools make possible a practice in which nature is encountered not as a power to which human beings are subject, but as *useful:* "The forest is a forest of timber, the mountain a quarry of rock, the river is a water power, the wind is wind 'in the sails'" (66). Second, thanks to their "in order to," tools refer to what is made with them, and at the same time to their future user: "The work is cut to his figure; he 'is' there as the work emerges" (66). Third, the work does not merely remain "in the domestic world of the workshop," but plays a role in "the public world." "In taking care of things, nature is discovered as having some definite direction on paths, streets, bridges, and buildings. A covered railroad platform takes bad weather into account, public lighting systems take darkness into account, the specific change of the presence and absence of daylight, the 'position of the sun'" (66).

Things, in short, disclose a world. When somebody uses a tool or piece of equipment, a referential structure comes about in which the object produced, the material out of which it is made, the future user, and the environment in which it has a place are related to each other. But that this is so, according to Heidegger, generally appears only when a handy or ready-to-hand tool or piece of equipment breaks down. When this happens, the tool suddenly demands attention for itself. The reliable dealings we are used to having with the tool are ruptured, and instead of withdrawing from our attention the tool suddenly forces itself upon us. Someone sits at a word processor focused on the text at hand and all of a sudden the computer freezes. The trustworthy world that developed around the computer—the open books, the keyboard, the screen, the cup of coffee; in short, the entire mutually referring network that Heidegger calls a world—is abruptly destroyed. The computer changes over from being one of the handy or ready-to-hand objects that shape this world to what Heidegger calls something *vorhanden:* "objectively present" in the newer translation, or "present-at-hand" in the older.

Its transparency is transformed into opacity. The computer no longer can be conveniently utilized in the practice of writing, but abruptly demands interaction with itself. The relation with the world around the computer that took place "through" it is disturbed. Only when it starts up again and everything works without a hitch is the world that was destroyed again restored.

When such a rupture takes place in the referential structure of the world that is disclosed by handy or ready-to-hand equipment, this structure itself becomes visible. Only when readiness-to-hand changes over into presence-at-hand does it become visible what takes place on the basis of readiness-to-hand; as soon as readiness-to-hand is disturbed, the references call attention to themselves. The coming to presence-at-hand of tools and equipment makes explicit their ability to be ready-to-hand and thus to make the world accessible for concernful, everyday dealings with it.

Seen in light of Heidegger's eventual philosophy of technology, this analysis of equipment is remarkable, to say the least. In *Being and Time*, technology, conceived in terms of equipment, is a way of *revealing* the world instead of a *reduction* of our access to it. "According to our foregoing interpretation, being-in-the-world signifies the unthematic, circumspect absorption in the references constitutive for the handiness [readiness-to-hand] of the totality of useful things" (71). As I shall make clear in the chapters to come, with this analysis Heidegger provides the first building blocks for a more satisfactory approach to technology. For he understands technology here in terms of concrete artifacts rather than by reducing it to its conditions, and also makes clear how these artifacts allow human beings to have access to reality.

The opposition between the early and late Heidegger that comes to light here has everything to do with the turn in his thinking. Beforehand, in order to understand being, Heidegger applied himself to an analysis of human existence in its relations to beings and its own being; afterward, to the history and self-manifestation of being itself. This opposition recurs in his thinking about technology. While the later Heidegger described technology from the perspective of the history of being and saw technological machines and devices as indices of this history, the early Heidegger addressed himself to an (ahistorical) analysis of the role of technology in the relation between human beings and their world. While the late Heidegger reduced technology (transcendentally) to the history of being, the early Heidegger approached it in terms of concrete artifacts.

But can the early Heidegger indeed be contrasted so sharply with the late? Heidegger himself always insisted that his turn cannot be conceived of as a break, that the two "phases" of his work are intertwined with each other and do not conflict.[16] Must this continuity, then, not also be present in his philosophy of technology? According to Hubert Dreyfus this is in fact the case.[17] Dreyfus points out the great temptation to think that, in *Being and Time,* Heidegger outlines a thoroughly technological picture of human being, especially given his claim that the wood is a forest of timber, the mountain a quarry of rock, and the river a source of water power. For the late Heidegger, the disclosure of a river as a source of water power amounts to a "setting upon" and ordering of nature as an energy source, à la the hydroelectric plant. Moreover, the involvement of human beings in the referential network of instruments is reminiscent of what Heidegger will later call the *Gestell.*

Dreyfus, however, believes that this kind of technological interpretation of *Being and Time* would be too simplistic. The reason that Heidegger analyzes tools and equipment therein is to make clear that the everyday dealing with things is more primordial than the scientific and abstract relation to them. And in contrast to the simple tools about which Heidegger speaks in *Being and Time,* modern technological artifacts depend precisely on a scientific and abstract disclosure of reality. According to Dreyfus, the most one can say is that Heidegger's picture of technology in *Being and Time* is ambiguous, and lies midway between ancient *technè* and modern technology.

Dreyfus resolves this ambiguity, and the tension that it creates between *Being and Time* and "The Question Concerning Technology," by historicizing *Being and Time.* Even though Heidegger holds that it is one (ahistorical) characteristic of equipment that it can be used, Dreyfus maintains that the meaning of "usefulness" is historically determined. This historical perspective, according to him, makes it possible to see *Being and Time* as belonging to a phase that lies between two periods of technology that Heidegger later distinguished explicitly. While Heidegger, in "The Question Concerning Technology," draws a sharp contrast between the craft-like *technè* of the ancient Greeks and the modern technology of our time, in *Being and Time,* according to Dreyfus, Heidegger is thinking of a stage midway between these two: that of industrial production. Heidegger characterizes the technology of the ancient Greeks as "bringing to presence" and modern technology as "setting-upon" and "ordering," but in *Being and Time* he speaks about equipment in terms of its *function,* its "in order to." This functionalization of equipment, according to Dreyfus, can be seen as a necessary intermediary step in the changeover from craft-like to modern technology.

16. See Heidegger's letter to William J. Richardson, S.J., which is reprinted as the foreword in Richardson (1963, ix–xxiii).

17. See Dreyfus (1992, 173–85).

The intermediary position of *Being and Time* is also evident to Dreyfus in the way in which Heidegger says that equipment discloses the world: while for craft-like technology nature was still *physis,* something that brought itself to presence and to which *technè* was subordinate, and while for modern technology nature was standing-reserve, Heidegger speaks of it in *Being and Time* as "raw material." Without it becoming standing-reserve, a fully stocked warehouse for whatever we want to do, nature nevertheless offers material that can be approached in a functional way in order to manufacture products.

However plausible it may seem to discover a middle step between *technè* and *Gestell,* it would be wrong to follow Dreyfus fully here, for he ignores an important and innovative aspect of Heidegger's analysis of tools and equipment: the attempt to articulate the way in which tools play a constitutive role in the relation between human beings and world. That reality was disclosed to the ancient Greek craftsman differently than within the age of industrial technology and within the modern-technological epoch does not affect the fact that human beings in each of these epochs employed tools and devices that shaped their relation to reality. While Heidegger's claim that nature is revealed in dealings with tools as a forest of timber or as a source of water power might be specifically linked to a historical phase of technology, his analysis of presence-at-hand and readiness-to-hand is not. The ready-to-hand way of being of equipment is not restricted to the technology of industrial production; the hammer of the ancient Greek silversmith and the word processor with which I in the twenty-first century write this book are ready-to-hand when they are in use, and disclose a world. In *Being and Time,* to be sure, Heidegger does not refer to word processors and uses instead the example of the hammer, and to that extent he does not explicitly link readiness-to-hand to modern technology. But the fact that he uses a traditional technical object like a hammer as an example reveals that his analysis is not specifically directed at industrial production either, as Dreyfus suggests.

In order to continue Heidegger's promising analysis of technological equipment, I shall pursue a route alternate to the one Dreyfus takes. My objection is not that Heidegger thinks the historicity of being too little, but that he does so too much. Just at the time of the *Kehre,* when Heidegger began to think radically about the historicity of being, he began to lose sight of specific technologies, as I pointed out in my critique of his later philosophy of technology. He then reduced technology to the meaning of being of which it was an expression, rather than approaching it in terms of specific technologies that coshape the relation between human beings and their world. In *Being and Time,* however, Heidegger sought to understand the role that specific technologies played in the relations between human beings and reality—an approach that, as I shall show, provides a point of departure for a "philosophy of technological artifacts."[18]

Equipment: Between Thing and Artwork

Heidegger's essay "The Origin of the Work of Art" marks the cautious beginning of what would become his *Kehre.* Here, too, he devotes considerable attention to tools and equipment, though he approaches them not with an eye to their role in praxis, but rather to the way in which they differ from two other kinds of objects, the "mere thing" (*das bloße Ding*) and the artwork.

> A piece of equipment, a pair of shoes for instance, when finished, is also self-contained like the mere thing, but it does not have the character of having taken shape by itself like the granite boulder. On the other hand, equipment displays an affinity with the art work insofar as it is something produced by the human hand. However, by its self-sufficient presence the work of art is similar rather to the mere thing, which has taken shape by itself and is self-contained. Nevertheless we do not count such works among mere things. As a rule it is the use-objects around us that are the nearest and authentic things. Thus the piece of equipment is half thing, because characterized by thingliness, and yet it is something more; at the same time it is half art work and yet something less, because lacking the self-sufficiency of the art work. Equipment has a peculiar position intermediate between thing and work, assuming that such a calculated ordering of them is permissible. (Heidegger 1971a, 29)

Heidegger conceives the piece of equipment here no longer in relation to the human beings who engage with it, but in relation to other objects. On the one hand it, like the "mere thing," is an object, though an object that is produced by

18. The way in which I draw the distinction here between ahistorical and historical aspects of *technology* differs from the way in which I used that distinction in the previous paragraph with reference to Heidegger's approach to *being.* In the case of being, the issue was that Heidegger was selective in applying the two aspects of being to traditional and modern technology. In the case of technology, I object to his (transcendentalist) reduction of technology to a historical phase in the history of being, for that overlooks how specific technologies always (ahistorically) play a role in the relation between human beings and their world, whether or not they are a manifestation of an always historical meaning of being.

human beings. On the other hand it, like the artwork, is produced by human beings, but it contrasts with an artwork because it requires a context of use in order to be genuinely present as a piece of equipment.

Because of this intermediate position, and because tools and equipment are the things that are closest and most familiar to human beings, Heidegger finds that equipment lies at the origin of the concept of "thinghood" found throughout the history of Western thought. The dominant way in which the thing has been conceived is via the "form-matter" structure: matter that has a particular form. This conception of things, Heidegger holds, is modeled after the tool, for it conceives the thing in terms of making, with tools and equipment being the objects that can be used for this making. Moreover, pieces of equipment themselves are the products of a making process in which matter is brought in a specific form. "As determinations of beings, accordingly, matter and form have their proper place in the essential nature of equipment" (28).

For Heidegger, however, the conception of the thing as "formed matter" is too limited.[19] From this conception—with equipment being the medium that imposes form on matter—the "mere thing" can only be conceived as matter with a form that was not explicitly intended by human beings and that is not usable. Meanwhile, the artwork appears as a piece of matter that was intentionally given a specific form, and that is not usable either but adds to its pure thingliness an aesthetic value. Heidegger finds the conceptual framework of matter versus form to be "hackneyed," applicable to anything and everything, and in a position neither to elucidate the thingly character of the thing, nor to distinguish between the different types of things: mere things, equipment, and works of art. What makes a thing a thing, according to Heidegger, is to be sought in its way of being, the way it is present to human beings in their relation to it. When the thing is approached in this way, then mere things, equipment, and works of art can be distinguished from each other by the different ways they are present to human beings.

Within this perspective, what, then, is the equipmental character of equipment? First, it should be made clear that by equipment, Heidegger means more

19. Heidegger also criticizes two other "interpretations of the thingness of the thing," namely, the approaches that treat the thing as a "bearer of properties" and as a "unity of a manifold of what is given in the senses." When the thing is conceived as the bearer of properties, according to Heidegger, what goes wrong is that the thing must first be present as a thing in order to discern its properties. This definition, that is, already presupposes a conception of the thing and therefore cannot help us. When the thing is conceived as a unity of a manifold of what is given "in the senses," what goes wrong is that we do not perceive "sensory impressions" but the things themselves—when we hear a door slam, for instance, we do not hear "sound" or "acoustical impressions" but a door slamming (Heidegger 1971c, 22–32).

than tools, but useful objects broadly interpreted—anything used to accomplish something. This much is clear, among other things, from his example in "The Origin of the Work of Art" of a pair of shoes rather than a conventional tool such as the famous hammer from *Being and Time.* His question is therefore: What makes a useful object a useful object? In *Being and Time,* he observed that a useful object is present as such when it withdraws from our attention in favor of the work being accomplished. To this, Heidegger now adds that a useful object can only be useful when it is *reliable.* When it wears out—when, for instance, the soles of a shoe wear away—the useful object loses its reliability, and therefore its usefulness. It changes over into a mere thing. According to Heidegger, therefore, reliability is the way of being of equipment.

This formulation can be seen as the positive articulation of the same phenomenon that was only negatively expressed in *Being and Time.* While in *Being and Time,* readiness-to-hand as the way of being of the tool or piece of equipment only became visible in negative situations—when the tool broke and thus became present-at-hand—Heidegger now gives this a positive content. In "The Origin of the Work of Art," the reliability of the useful thing is what gives human beings access to the world. The peasant's shoes, which Heidegger here uses as an example, reveal her world to her. Thanks to them, she can plod over the field, till the soil, wend her way home—and therefore give a measure of certainty to her daily bread. Useful objects give human beings an access to being and thus shape the world that gathers around them.

After tools and equipment, Heidegger introduces another class of objects that can gather a world: artworks. An artwork "is" in a completely different way than a piece of equipment. In contrast to the piece of equipment, which can only be present as equipment when used, the artwork rests within itself. In this self-containment, according to Heidegger, it is able to bring beings into unconcealment to disclose a world. Here he appeals, connecting to the example above, to a painting by Van Gogh of a pair of peasant shoes. When we look at the painting, we are "suddenly somewhere else than we usually tend to be" (35). The painting lets the shoes be present, reveals them, brings them into unconcealment. The artwork therefore belongs to the domain of *alètheia,* which Heidegger translates as "truth," as became clear above: "The art work lets us know what shoes are in truth" (35). Art, according to Heidegger, should not only be conceived in terms of beauty or of the representation of reality, but must be viewed first and foremost as "the truth of beings setting itself to work" (36). What an artwork does is to let beings come to presence, to disclose. "The art work opens up in its own way the Being of beings" (39).

Through this analysis of the artwork as the disclosing of beings, as a "letting come to presence," the piece of equipment comes to have a different status from what it had in *Being and Time.* While Heidegger initially conceived the coming about of a world from out of equipment, he conceives of *alètheia* as more fundamental in "The Origin of the Work of Art." As Pöggeler writes, "If the world is seen in terms of equipment, then it shows itself as 'environment,' as the context of references and significance of the ready-to-hand" (Pöggeler 1991, 168). This context of references is linked with the praxis of human beings thanks to the "in-order-to," the "for-the-sake-of" of the piece of equipment. But "the for-the-sake-of in which the world is grounded . . . is thought one-sidedly in terms of the understanding which projects significance" (168). In *Being and Time,* according to Pöggeler, Heidegger thinks in too "innerworldly" a fashion; he only allows the world to show itself through human activity, and does not yet take into account the event of being by which beings come to presence (169). In "The Origin of the Work of Art," Heidegger links this dimension of the coming to presence of beings, which precedes the disclosure of beings by equipment, with the work of art. For there to be beings for us in the context of significance of a world, these beings have to be unconcealed, disclosed—which is precisely what a work of art does.

If the world is always a result of a *disclosure* of beings, of unconcealment, "something" must precede this world that can be disclosed as world. For this reason, in "The Origin of the Work of Art," Heidegger introduces the concept of "earth" alongside that of "world." "Earth" consists, as William Richardson puts it, of "the material elements (e.g., pigments, marble, musical notes) out of which the work is fashioned" (Richardson 1963, 406). The environment of the artwork—what is disclosed as world by it—also belongs to the earth. In the earth, the dimension of concealment is localized, from which beings can show themselves in a particular way as they emerge into the unconcealment of a world. Heidegger illustrates these thoughts through the aforementioned example of a Greek temple. The temple lets its environment appear in a particular way. Only in the presence of the temple can what is concealed in the earth emerge into unconcealment in a particular way—the rock is disclosed as that which is able to support the temple, the storm as something the temple is able to face, and so forth. Artworks not only open up a context of significance, but also set into effect the "coming into being" of a world.

Heidegger conceives the artwork as the setting into work of a "struggle" between world and earth; it is shaped out of the concealment of the earth, and discloses this earth into a world. Heidegger speaks of a "struggle" here, because on the one hand earth is the *concealed* from out of which the work is made, while at the same time it is *revealed* in a specific way, since the artwork brings it into the unconcealment of a world. In this struggle, truth, conceived as unconcealment (*alètheia*), "happens." As Pöggeler puts it: "The structure of truth as unconcealment is the togetherness of world and earth. World is the articulation of the open, the clearing, earth the articulation of the self-closing, of the concealment as the sheltering. World is never without earth. It is not simply the clearing, but rather the clearing which comes forth from unconcealment" (Pöggeler 1991, 172).

In "The Origin of the Work of Art," the dimension of being is linked in a second way with artworks and denied to equipment. For Heidegger tries to conceive the artwork and the piece of equipment not only in terms of objects, but also in terms of the way in which they come about; by the way they are brought forth. Heidegger calls the process by which artworks are brought forth "creation," and the process by which equipment is brought forth "making." He asks himself, "But what is it that distinguishes bringing forth as creation from bringing forth in the mode of making?" (Heidegger 1971a, 58). To answer that question, Heidegger also places the bringing forth of objects in the domain of *alètheia*. As he will do later in "The Question Concerning Technology," he appeals to the ancient Greeks, who indicated "making" with the word "*technè*," which had the double meaning of "handicraft" and "artwork." *Technè* does not designate "a kind of practical performance," but rather designates "a mode of knowing." He continues, "To know means to have seen, in the widest sense of seeing, which means to apprehend what is present as such. . . . *Technè*, as knowledge experienced in the Greek manner, is a bringing forth of beings in that it *brings forth* present beings as such beings *out of* concealedness and specifically *into* the unconcealedness of their appearance; *technè* never signifies the action of making" (59). Bringing forth thus means to allow beings to become present. It is a form of disclosing, and therefore plays out in the domain of *alètheia*. Not only does the artwork *itself* evoke the happening of becoming unconcealed, so does its *creating*.

According to Heidegger, important differences exist between the creation of an artwork and the production of equipment. The production of equipment lacks two important characteristics possessed by the creation of a work of art. The first is that only in the creation of an artwork is the struggle between world and earth set into work. The artwork opens a world and can do so only thanks to the fact that it is created "out" of something; that is, earth. This dimension of concealment that belongs to the earth, however, disappears in the production of equipment, for the material out of which the equipment is made appears only when the equipment breaks. The production of equipment does not *show* the earth, but rather *uses it up.* While a work of art shows the "something" out of which it was made—by disclosing that "something" and bringing it into unconcealment—equipment that is used withdraws from human attention in order to make possible the practical activity for which it is used. The piece of equipment does not attract attention to itself or toward that out of which it is made, but rather disappears in usefulness.

The second difference between the creation of an artwork and the production of equipment is that for the artwork "createdness is expressly created into the created being," while the production of equipment is not prominent in the equipment itself. The artwork shows that its creation was a happening of becoming unconcealed, or "that such a work *is* at all rather than is not" (65). Such is not the case with the piece of equipment, once again because it "disappears in usefulness. The more handy (ready-to-hand) a piece of equipment is, the more inconspicuous it remains that, for example, such a hammer is and the more exclusively does the equipment keep itself in its equipmentality" (70).

"The Origin of the Work of Art" announces Heidegger's turn toward being. He gradually develops a new way of thinking about beings, and therefore about technological equipment and artifacts. First of all, he no longer concerns himself with the contexts of meaning that crystallize around equipment, but rather with the role of objects in the happening of being. Second, he begins to think about objects in terms of the way of *bringing forth* by which objects come into being. These are the first key steps toward Heidegger's later transcendentalism. He begins to question no longer "forward," into the world which exists around an object, but now "backward," into the conditions of possibility of the origin of the world and the object; their "coming into being." The emerging priority of that coming into being makes progressively less relevant the world-disclosing capability of tools and equipment.

But these are only the initial steps. Heidegger's direction of thinking is beginning to turn backward, but "The Origin of the Work of Art" does not yet absolutize this new perspective. Specific technologies, in the form of equipment, continue to play a role in the coming about of the contexts of meaning of a world. And Heidegger's analysis of the Greek temple explicitly shows how a specific artifact can call a world into appearance and thereby make visible the happening of becoming unconcealed. In this transitional work, Heidegger adds another dimension to his analysis of equipment in *Being and Time,* without shutting off his earlier perspective.

The Thing Things

"Building, Dwelling, Thinking" and "The Thing" form the next step in the development of Heidegger's understanding of things, where he thinks ever more firmly from the perspective of being and ever less from the manner in which things shape the relation between human beings and their world. In these writings, what appears decisive to Heidegger is the ability of things to "reveal" or "let come into being." Things are no longer investigated for the ways in which they make practices possible and thereby disclose a world. The ability to "let come into being," that was reserved in "The Origin of the Work of Art" only for artworks, Heidegger now attributes to all things. And while there he made it clear only *that* beings were "brought into being" by a thing, he is now concerned to make clear *how* they come into being. He no longer draws a distinction between "mere things," equipment, and artworks, but speaks comprehensively about "the thing." From his examples, however, it is clear that he has useful things foremost in mind.

"The thing," according to Heidegger, is in a position to let a world come into being, and indeed in a specific way. He poses to himself the question of "what a thing is as thing." His answer appears at first glance cryptic enough: he says that "the thing things" (Heidegger 1971c, 174). With this conjunction of a noun with a verb made from it, Heidegger means to express that what a thing does, the way in which a thing is present as thing, cannot be reduced to something non-thingly and must be conceived from the thing itself. What, then, is the "thinging" of a thing? Heidegger finds this in the "gathering of the fourfold." The thing calls forth a world by opening a space for the earth and sky, divinities and mortals. In "Building, Dwelling, Thinking," as already mentioned, Heidegger appeals to the example of a bridge: "The bridge *gathers* the earth as landscape around the stream," its arches "ready for the sky's weather and its fickle nature," escorting "the lingering and hastening ways" of mortals as they go about their business, but always "before the divinities," and in a way that can allow us to "think of, and visibly give thanks for, their presence" (152–53).

In "The Thing," Heidegger explores how a jug gathers the fourfold. In order to comprehend the jug in its jug-character, Heidegger finds that he has to think about the way in which it is present as a jug—as an object with which a drink can be poured out. "The jug's jug-character consists in the poured gift of the pouring out." In the drink that is poured out stays "the earth, which receives the rain and dew of the sky," by which the drink "is a drink for the mortals," but also can be "a libation poured out for the immortal gods" (172):

> The gift of the outpouring is a gift because it stays earth and sky, divinities and mortals. Yet staying is now no longer the mere persisting of something that is here. Staying appropriates. It brings the four into the light of their mutual belonging. From out of the staying's simple onefoldness they are betrothed, entrusted to one another. At one in thus being entrusted to one another, they are unconcealed. The gift of the outpouring stays the onefold of the fourfold of the four. And in the poured gift the jug presences as jug. . . . What is gathered in the gift gathers itself in appropriatively staying the fourfold. (173–74)

In its "thinging," according to Heidegger, the thing discloses a world ("the thing things world")—a world that is experienced in terms of earth and heaven, mortals and gods. This "thinging" of the thing now becomes the measure of what a thing is. "The jug is a thing, insofar as it things. The presence of something present such as the jug comes into its own, appropriatively manifests and determines itself, only from the thinging of the thing" (177).

The "thinging" thing discloses a world in a completely different way than did the piece of equipment in *Being and Time.* It does not shape the relation between humans and their world by mediating praxis, but rather sets to work the "appropriating" or "happening" of the "staying of the fourfold." The thing continues to have a place in praxis: a jug things when it is used to pour, a bridge when it is used to cross over a river. But the question of what active role the thing has in the praxis that surrounds it has vanished from Heidegger's view. Just as in *Being and Time* and "The Origin of the Work of Art," the thing is that around which a world forms, but what is essential about this world-disclosing capacity in "The Thing" is no longer thought of as arising from the specific way in which the thing provides access to beings. The disclosure of a world is now thought of as arising from its "becoming unconcealed"; Heidegger is concerned with the *coming to presence* of beings, which manifests itself through things and not with the *specific role of actual things* in that process. While in *Being and Time* each piece of equipment disclosed a world in its own specific way, in "The Thing" the way in which a thing discloses a world is already predetermined: all things disclose a world in terms of the fourfold.

Apart from the question of how persuasive it is that things disclose the world in terms of sky, earth, gods, and mortals, it is clear that, despite the title of the work, in "The Thing" Heidegger begins to lose sight of things themselves. While Heidegger's speaking about the "thinging" of the thing suggests that he is trying to develop a way of thinking about things that does not reduce them to something non-thingly, he does not finally succeed in so doing. In the "letting be unconcealed of the fourfold," things play at most a facilitating role, and do not actually contribute anything to the way in which the world is disclosed. Whether or not we humans use a bridge or a jug, the world is still disclosed in terms of heaven and earth, divinities and mortals. In the end, Heidegger appears to be interested not in *things,* but rather in the way in which being can show itself. The "thinging" of the thing comes down to "letting come into being," and therefore what is essential about the thing is, remarkably enough, something that is unconnected with the specific thing human beings are dealing with—the happening of being. The active role of the thing itself in world disclosure, which was thought in *Being and Time* in terms of the specific networks of meaning that crystallize around a concrete thing, has now disappeared.

This disappearance of the active role of specific things is illustrated by the fact that the world-disclosed things do not anymore bear the stamp of the specific way in which they disclose it, but rather of the way of thinking of human beings. This way of thinking can either stand open for being, which announces itself by the thing, or be blind to it. According to Heidegger, the thing needs a certain receptivity from human beings in order to be present as a *thing.* A part of this receptivity is, for Heidegger, the *andenkende denken,* what might be called meditative thinking, which involves a "step back from the thinking that merely represents—that is, explains—to the thinking that responds and recalls" (181). This kind of thinking not only thinks beings but also their being. It consists of a nearness to things that in our time is usually lacking. Nowadays, according to Heidegger, there is no nearness but rather "distancelessness." Thanks to media and modern transportation, distances are becoming shorter, but this leads, not to nearness, but only to uniformity (165–66).

The "letting come into being," through which Heidegger only understood the manner of presencing of a work of art in "The Origin of the Work of Art," becomes in "The Thing" and "Building, Dwelling, Thinking" the way to understand things in general and useful things in particular. A useful thing no longer discloses a world by shaping human praxis and thus the relation between human beings and their world, but lets the world come to being as fourfold. The thing, therefore, becomes no more than an intermediary, a spot where being does its work, without the thing itself playing an active role in the way in which a world comes to presence.

From Things to Being

"The Question Concerning Technology" constitutes the final step in the transition in Heidegger's thought from things to being that announces itself through things.

Heidegger now *reduces* things to being by regarding them as elements in the history of being. Whereas he came to understand useful things in terms of "coming into being" in "The Thing"—which he only did with regard to works of art in "The Origin of the Work of Art"—he now historicizes this very "coming into being." For in "The Question Concerning Technology" Heidegger adds a further dimension to the "coming into being": it happens always inside a specific way of revealing, a historical "sending" of being. The ancient *technè* of the craftsman and the modern technology of our present age are the products of different epochs of the history of being; they represent different interpretations of what it means "to be."

In this way Heidegger's "backward" thinking becomes absolutized. Technological artifacts no longer are the *source* of the disclosure of reality, but instead *spring from* it. Things no longer play an active role in the way in which human beings have access to reality, as the piece of equipment did in *Being and Time* and "The Origin of the Work of Art," but instead merely express an always historical meaning of being. Heidegger does not think things as such, but reduces them to the way of unconcealment that prevails in a particular historical epoch. He describes the disclosure of crafted tools and modern technological devices alike in terms of their conditions of possibility; the ways of unconcealment that precede them. Heidegger sees the hydroelectric plant, which discloses the Rhine as an energy source, not as the *origin* but as the *consequence* of "setting-upon" or "ordering." The Rhine is ultimately not disclosed as an energy source *by* the power plant, but because it cannot be disclosed otherwise within the prevailing way in which beings in our epoch "are." It is not the machines that disclose beings as standing-reserve; rather, the machines exist only because beings are *already present* as standing-reserve.

What a work of art was able to do, according to "The Origin of the Work of Art," a technological device cannot do: let beings emerge out of concealment into unconcealment. The reason for this is not the differences between the way in which art works and machines disclose reality, but the prevailing conception of being. The being of the Rhine, as disclosed by a hydroelectric plant, is forgotten, while the being of the temple environment "happens" precisely because of the temple itself. What prevents a power plant from letting its environment come to presence in just the way a temple does? The question calls for an answer, but from Heidegger's perspective in "The Question Concerning Technology" it is no longer relevant. Heidegger, after all, insists that in the epoch of machines being takes such a form that it is not understood anymore as "coming to presence" but only as a product of human activity. Heidegger now thinks radically from the perspective of being, and technologies themselves ultimately have no role to play in such an approach.

Meanwhile, Heidegger thinks not only modern technological devices from the perspective of being, but also traditional technical objects like tools and pieces of equipment. He creates a contrast between *technè,* which, with the help of equipment, brings beings out of concealment into unconcealment, and modern technology, which, with the help of machines, sets upon and orders beings. In *Being and Time* tools and pieces of equipment were still the center around which a world crystallizes. In "The Origin of the Work of Art" Heidegger continued to conceive them in this way, but he also contrasted them with artworks in order to make clear that tools and equipment lack the ability of the artwork to "let come into being." In "The Question Concerning Technology," however, Heidegger *does* ascribe this very ability to tools and pieces of equipment, at least those of the ancient Greeks. In passing, therefore, Heidegger revises his conception of the tool, and only thanks to this revision could modern technology be viewed as degenerate. The *technè* of the ancient Greek silversmith, which cannot be thought without the help of the tool, according to Heidegger always consisted of the "letting come to presence" of his products, such as silver chalices. *Technè* therefore belonged to the domain of *alètheia,* of letting-come-to-being—in contrast to modern technology, which does not trace the origin of its products to being but rather to its own manipulative activities.

What happens in "The Question Concerning Technology," in short, is that technology is wholly placed in the context of the history of being, and is no longer viewed in terms of specific technological artifacts. Heidegger makes this explicit by the remark that "the essence of technology" is not to be found among things like "rods, pistons, and chassis." "The Question Concerning Technology" thus addresses itself to technology not in the sense of "technological objects" but in the sense of "producing." It is precisely this that makes possible a connection with being, for in production a thing comes to "be." Production is therefore the outstanding point of contact with which to place technology in the context of being and to make visible how objects come into being in different ways.

That Heidegger approaches technology here principally in terms of making or producing and not in terms of objects does not imply that the criticism that he does not take specific technologies sufficiently seriously is stating the obvious. For Heidegger does indeed devote attention to specific technologies—as when he speaks about hydroelectric plants as the consequence of the setting-upon of energy—but he does so in a way that causes them to drop out of sight again, reduced to outcomes of the history of being.

In "The Memorial Address" this becomes still more clear. There, in contrast to his usage in "The Question Concerning Technology," Heidegger frequently speaks about "technological objects." The comportment that he calls "releasement" is explicitly concerned with such objects. Modern devices "claim, enchain, drag along, press and impose upon" us. A comportment of releasement frees us from their grasp, and lets them be "as things which are nothing absolute but remain dependent upon something higher." At the same time, it allows us to notice that "while the production and use of machines demands of us another relation to things, it is not a meaningless relation" (Heidegger 1966, 54). Reflections on the way in which technological devices disclose reality are thus intertwined with the idea that this way of disclosure is "dependent on something higher."

But however much Heidegger tries—in a "forward" fashion—to describe the way that devices offer another, and specific, way of disclosing reality, one that requires a different kind of comportment from human beings, he cannot avoid—in a "backward" fashion—to appeal to an a priori "sending of being" as the origin of this way of disclosing. Heidegger does not deny that technological devices in fact disclose reality in a specific way, but his account is inadequate because he thinks this disclosure ultimately from the point of view of the history of being and not from the technological devices themselves. After all, the examples I gave above of the ways in which certain modern technologies can disclose reality other than as standing-reserve, together with Heidegger's own analysis of tools and equipment in *Being and Time,* reveal how insufficient the picture of technology is that emerges from the works of the late Heidegger. Heidegger's transcendentalist approach is not able to give an adequate account of modern, technological artifacts.

It can be concluded that Heidegger's concern for things continued to disappear from his work as it continued to progress. In contrast to what is suggested by his own formula that "the thing things," he appears to have turned away from things in the course of his thinking. However much Heidegger may have thought about objects, ultimately he reduced them, by backward thinking, to the being that announces itself through them. Only his early work offers a connection, thinking forward, to answer the question of the role that things play in the way in which reality can be present to human beings. I shall be building on these analyses for the remainder of this book.

CONCLUSION

In Heidegger's philosophy of technology, technology can be regarded neither as a means nor as a human activity, but rather must be regarded as a way of disclosing reality. The way in which technologies disclose reality, however, is determined by the way in which reality can "be" in the modern age, since the ruling understanding of "being" determines what "counts as" reality. In the epoch of modern technology, according to Heidegger, reality is disclosed as a standing-reserve of raw material and energy available for our use—not because we humans will it to be that way, but because this is the way the world manifests itself to us. This dominating way of reality disclosure leads humans to forget being itself; that things come to presence and always show themselves to us in a particular way.

To retrieve openness to being, according to Heidegger, humans have to await the arrival of a new way of being. This becomes possible when human beings reflect on technology, for it then becomes clear that even in the epoch of modern technology things come to presence in a particular way, even and especially when humans think that all beings are their own products. The coming to presence of beings shows itself most clearly in the realm of art, for an artist's work involves coming to presence par excellence. A condition for the emergence of a new way of being is that humans neither hold themselves apart from technology, nor fully give themselves over to it. It requires, rather, an attitude of "releasement" vis-à-vis technological artifacts, in which humans use them without letting them fully determine their relation to the world. Such releasement preserves the space to be open for the coming into being of entities and for the possibility of a new way of being, in which being itself is no longer concealed.

The most important critique to be made of this approach is that Heidegger does not fully succeed in developing an adequate hermeneutical perspective on technology, for he reduces technology to its conditions of possibility and thereby fails to connect with specific technologies. Such a transcendentalist approach, however, appears to be characteristic only of his later thought, beginning with the period of the *Kehre.* The early Heidegger, in *Being and Time,* does provide a substantial foundation for a hermeneutical philosophy of technology that analyzes the role that actual technologies play in the way in which reality acquires meaning for human beings. For in *Being and Time* he does not see technological artifacts as manifestations of a sending of being, but instead tries to investigate how actual tools and equipment constitute networks of meaning.

These early initiatives by Heidegger therefore can provide the foundations for a phenomenological philosophy of technology that takes actual technologies seriously. Only by thinking from actual technological artifacts can a philosophy of technology be developed that genuinely addresses *technology* instead of its *conditions.* Such an approach would do justice to the thought contained in the phrase, "the thing things," for only then would the thing not be reduced to something non-thingly but thought *as thing.* In the subsequent chapters of this book I shall inquire how to build on these thoughts to develop a "philosophy of technological artifacts."

PART II

Philosophy From Things

3

Postphenomonology

INTRODUCTION

How, then, to overcome the limitations of classical philosophy of technology? These limitations, as we have seen, are all too clear in the philosophies of Jaspers and Heidegger. Technology is primarily conceived as a form of alienation: it alienates human beings from themselves in preventing them from achieving authentic existence, and it alienates human beings from the world in denying them a meaningful place to exist. This negative judgment can, in part at least, be related to the historical situation in which Jaspers and Heidegger formulated their thought. In the first half of the twentieth century, society was undergoing rapid changes thanks to the influence of industrialization. The repetitive, monotonous character of assembly-line work appeared to herald a new kind of mass society and homogenized existence; cold, anonymous industrial complexes seemed to indicate the onset of a reduced relation to the world. But the classical diagnosis appeared to be premature, and failed to foresee the ways in which technological society and culture would develop. Today, over half a century later, we see that humanity has not been entirely swallowed up inside the production apparatus, and is able to approach reality not exclusively as a storehouse of raw materials.

Close inspection revealed that Jaspers and Heidegger failed to support their analysis of technology adequately. They reduced technology to its conditions of possibility and then proceeded as if what they said about these conditions applied to technology itself. Heidegger's hermeneutical approach attempted to understand technology as an alienating way of disclosing reality, reducing concrete technological artifacts to the fruits of such disclosing. Jaspers's existential philosophy of technology attempted to understand technology in terms of bureaucracy, mass production, and the "limits of technology," and then likewise reduced it to what this made possible or to what imposed limits on it. Both philosophies appear to be governed by what one might call a "transcendental fix." In the style of transcendental philosophy, they tried to apprehend technology one-sidedly from its conditions of possibility. They thought "backward," reducing concrete technologies to nontechnological things such as "technological thinking" or "the system of mass production," with technology itself, in the end, falling out of the picture.

One of the most important counters to the standard classical picture of technology has come from empirical research into the development and use of technologies, which has revealed that this classical picture fails to match technological reality. The advance of technology does not follow a single dynamic but is rather the contingent outcome of a set of complex and interactive processes. But however much the empirical approach, contrary to Jaspers and Heidegger, gives concrete technologies the attention they deserve, it does not by itself present an adequate alternative to the classical philosophy of technology. For in empirical technology studies the hermeneutic and existential questions posed by the classical philosophers of technology fall out of the picture: What is the role that technology plays in human existence and in the relation between human beings and reality?

A full-fledged philosophy of technology would have to do justice to the concrete empirical reality of technology without giving up on the philosophical issues posed by the classical approach. In this chapter I shall show how to do so by shifting the philosophical attention from the conditions of technology to technology itself—to the technological devices and objects that are virtually ubiquitous in our daily lives—thereby seeking to understand them via the role that they play in our society and culture.[1] But how is it possible to think about

1. In *Thinking Through Technology,* Carl Mitcham distinguishes among four different manifestations through which technology can be analyzed: as knowledge, as activity, as "will," and as object. The word "technology" can indicate a form of *knowledge,* such as sensory-motor skills, rules of thumb, and technological theories. But it can also indicate the *activity* of design, manufacture, and use of new technologies; the (Heideggerian) technological *will to power;* or technological *artifacts.*

technology from the perspective of artifacts? This is the question to be addressed in this chapter, and the answer will give rise, in the succeeding chapters, to a conceptual framework that can form the basis of a philosophy of technological artifacts.

In order to sketch out the principles of such a philosophy, I shall begin by taking up a new interpretation of the phenomenological tradition from which Jaspers and Heidegger developed their positions. Following Don Ihde, I shall characterize this interpretation as "postphenomenology," but I shall give it a broader definition than he himself does. I shall sketch out the contours of a postphenomenological perspective on technology that is able to do justice to concrete technologies without abandoning the hermeneutical and existential questions that inspire it.

EMPIRICAL RESEARCH INTO TECHNOLOGY

The classical philosophical image of technology has received severe criticism from empirical research into technology. The chief weaknesses that this empirical research exposed were the overgeneralizations and false determinism in the image of technology offered by the classical philosophy of technology. Technology was supposed to develop autonomously, with society adapting in its wake—an image to which Jaspers and Heidegger could indeed subscribe. According to Jaspers, technology has become an end in itself, demonically holding society in its grip; according to Heidegger technology is a sending of being and our only hope is to await expectantly a new configuration of being to take shape.

Empirical research into the evolution of specific technologies and the reciprocal interactions that they have with society have undermined these deterministic approaches. This research took place initially under the social constructivist flag. Technology was conceived as the result of human activity instead of as something autonomous, paralleling an earlier development in science studies. Its evolution was viewed as an outcome of choices made by human beings in the specific circumstances in which they find themselves. The development of technologies was regarded as socially constructed, rather than following any innate pattern: each technology comes into being via a contingent process of social interaction.

This approach to technology became known via the acronym SCOT: the Social Construction of Technology. One of its important proponents was Wiebe

---

My concern here is with technology in the latter sense: concrete technological objects or artifacts (Mitcham 1994, 137–60).

Bijker, one of whose well-known studies concerned the coming about of the design of the modern bicycle as we now know it. Bijker examined other, past bicycle designs and how the interaction between these and the social context gave birth to the contemporary design. He thereby demonstrated that this cannot be viewed as the "one best solution," but rather as the outcome of a complex power struggle between a multiplicity of "relevant social groups" who wanted to use bicycles for specific ends, found particular models too dangerous, and so forth.[2]

But limitations were soon found in the SCOT approach as it became clear that the technologies themselves also played an active role in "social" interaction processes. The example of the microwave oven mentioned in the introduction is a beautiful illustration: the factors that determine whether human beings take their meals together include not just human beings but also the microwave itself. Reducing technology to social interactions therefore fails to do justice to the active role played by technologies themselves. Phrased in terms of my criticism of Jaspers and Heidegger in the previous chapters, it can also be stated that the empirical approach came to the conclusion that it could not do justice to technology by reducing it to its conditions.

The successor to social constructivism named itself simply "constructivism." The most influential framework for a constructivist approach to technology is supplied by the "actor-network" theory proposed by French philosopher and anthropologist Bruno Latour. Latour describes reality in terms of actors who link and interact with each other via networks. He calls his descriptions "symmetric," inasmuch as they do not make any a priori distinction between human and nonhuman actors. Things and artifacts, too, can become actors and thus deserve to be studied on a par with humans. Technologies do not merely arise from an interaction, but also play an active role in it. The speed with which we drive our cars, to use one of Latour's examples, is not only a function of our own choices and desires, but can also be affected by the existence of things like speed bumps on the road. The term "actor," in fact, misleads to the extent that it connotes human behavior; Latour therefore prefers to speak of "actants" rather than "actors."

Latour sees all phenomena that can possibly be encountered in the world—and especially technologies and scientific theories—as parts of networks of relations between actants. Some networks are vast, others small. The computer with which I write this book, for example, is part of an extensive network that includes software manufacturers, hardware manufacturers, the university where I work, which has given me access to the computer, the university's systems manager,

2. See Bijker (1995), chapter 2: "King of the Road: The Social Construction of the Safety Bicycle.

and the automobile of my colleague with which the computer was delivered to my home. Without that network my computer would not be available for me as a functioning device in my study. This network is usually hidden: the only thing that matters when I work is that the computer is in front of me and functional. To use Latour's language, the computer is through and through a black box; it is viewed as an independent, self-standing object with both its internal engineering and the relations with other things that make it work hidden from view. But if we want to understand how the computer came to be in my study, or if we want to fix it when it breaks, then the network of which it is a part suddenly comes to light—or at any rate a part of that network.

Latour's actor-network theory arose from his research into the coming about of scientific knowledge. According to him, scientific knowledge cannot be understood as "the truth" about "reality itself," but is a product of the interaction between humans and nonhumans in a network involving definitions, problem-setting, experiments, and observations. That network consists of relations between researchers and the phenomena into which they are inquiring, and everything that plays a role in those relations. Generally the outcome of this interaction is black-boxed, just as is my computer in the above example: human beings take the theory as obviously "true" and forget about all the efforts that the scientists had to take in piecing it together. But when the black box of such a theory is opened, its obviousness becomes far less obvious—for this brings to light the enormous amount of activity that was required in order to make the theory seem "true." A scientific theory must not be seen as a mirroring of reality, but as the product of a network of relations that link researchers with the phenomenon in question. Scientific knowledge is thus not discovered but constructed; it is an edifice that, up to a point, could have been otherwise.

This constructivist conception of reality can be used to investigate not only scientific knowledge but technology as well. Its advantage is that it does not simply reduce technologies to networks of social interactions, as did the social constructivist conception, but also analyzes the ways in which technologies themselves coshape the interactions. In Chapter 5 I shall work out in more detail the way in which Latour analyzes the active role of technologies. It will suffice here to present the conclusion that actor-network theory offers more than a "backward" approach to technology, but pays attention to what technology actually does in its context, without reducing its role solely to its origins.

Latour's empirical approach to technology, however, does not offer a true alternative to the classical philosophy of technology. The questions that classical philosophy of technology posed play hardly any role in Latour's work—neither the existential question of the role technology plays in human existence nor the hermeneutical question of how technology coshapes the access human beings have to reality. In order to provide the necessary answers to such phenomenological questions, a new interpretation of the phenomenological perspective itself needs to be worked out.

BEYOND CLASSICAL PHENOMENOLOGY

Someone who uses a phenomenological approach to technology in the twenty-first century still has some explaining to do. Phenomenology was an important tradition in Continental philosophy in the first half of the twentieth century, but its influence has waned. Its fundaments have been challenged as problematic due to a number of philosophical developments in the second half of the twentieth century, such as the linguistic turn and the subsequent appearance of postmodernism.[3] Phenomenology was thrust on the defensive in its response to these challenges, thanks to the suspicion that it requires recourse to an "authentic" or "original" access to reality. Its suppositions seem to mesh poorly with the contemporary emphasis on locality and context-dependence, according to which human access to reality is never direct but always mediated. In light of postmodernism and the linguistic turn, phenomenology seems to be obsolete, a romantic throwback. What could such a tradition still have to offer?

Yet phenomenology can be reinterpreted without the alienation thesis. It can be productively applied in a way that provides the framework for a "philosophy of artifacts." The suspicion that classical phenomenology misunderstands the locality and context-dependence of human knowledge is understandable when the context in which it developed itself is taken into account. Phenomenology presented itself—wrongly, as I shall make clear—as a philosophical method that sought to describe "reality itself." It had good reasons for so doing, which reveal how closely allied phenomenology is with postmodernism. For phenomenology opposes itself to the absolutization of the positivistic view of the world arising from modern natural science, which claims to describe reality as it actually is. Phenomenology sees this absolutization as going too far, inasmuch as it fails to disclose other aspects of reality that are not amenable to scientific analysis. In phenomenological terms, science reveals, not "reality itself," but a reduced reality.

3. Even Latour's actor-network theory can be counted among these developments, though in a somewhat headstrong way; he calls his thinking not postmodern but amodern (see Chapter 5).

Dutch psychiatrist-phenomenologist J. H. Van den Berg, for instance, speaks of a lived reality as opposed to the dismantled world provided by sciences:

> Have you ever drunk H_2O? Me neither. Nor do I want to. Real humans drink water. Have you ever gone swimming in H_2O + NaCl? What a shame! I've swum in the North Sea, the Atlantic Ocean, the Pacific, the Mississippi, the Po, and the Adriatic. That's an experience—a genuine experience. . . . Phenomenology attempts, once again, to bring to center stage this original and meaningful world, which, of course is always there, and, to some extent, to deny as well that the natural sciences are right. The natural sciences work with the mere skeletons of things. Better: the natural sciences work with the conditions of things.[4]

Phenomenology's protest against the absolutization of the scientific perspective is still timely, but its claim to provide access to genuine reality and its full significance is not. It correctly pointed out that the scientific disclosure of reality is not a disclosure of "reality itself" but always that of a quite specific kind—but from this fact it failed to draw the conclusion that no final contact with "true reality" is possible at all, and that therefore even "lived reality" is always lived in a specific way. This is the crucial step that needs to be made in light of postmodernism and the linguistic turn. The tree that I climb is real for me in a different way than the one whose cells and sap I study, but so is the tree that I photograph, chop down to use for firewood, or cut up to build a table. None of these disclosures can claim to reveal the "true" tree: they are each equally true.

Phenomenology's claim to regain access to an original, meaning-rich world, one lost by the natural sciences, makes its position difficult and open to challenge. It claimed to take its point of departure from an original position, one from which real human beings have become alienated, whereas among philosophers the insight grew that the human experience of reality is always mediated. The "original world, rich in meaning" of which Van den Berg speaks is thus just as mediated as the scientific world—by language, frameworks of interpretation, and social and behavioral contexts. Science, therefore, does not involve an exclusion of the meaning of the world, but a new and different kind of disclosure of it.[5]

4. This quotation is from a lecture Van den Berg gave in 1991 at the Vrije Universiteit Amsterdam, quoted in Heij (1995).

5. A second reason why phenomenology has fallen under suspicion is due to the so-called *Wesensschau* (essential intuition) of Husserl's methodology. Contrary to what is suggested by this term, the Husserlian *Wesensschau* does not seek to intuit "true" or "authentic" reality, but is rather

Against Method

Phenomenology, however, does not need to take shape as a philosophy of alienation. It originally took this direction in part in reaction to the positivistic worldview, but the ideas that lie at its foundation can be worked out in an entirely different manner. To see this, we need only direct our attention to these key ideas. Consider, for instance, the following passages from the famous preface to Maurice Merleau-Ponty's *Phenomenology of Perception*, which reveal not only the problematic aspects of classical phenomenology mentioned above but also the possibility of overcoming them.

> [P]henomenology can be practiced and identified as a manner or style of thinking. . . . It is a manner of describing, not of explaining or analyzing. Husserl's first directive to phenomenology, in its early stages, is to be a "descriptive psychology," or to return to the "things themselves." is from the start a rejection of science. . . . All my knowledge of the world, even my scientific knowledge, is gained from my own particular point of view, or from some experience of the world without which the symbols of science would be meaningless. The whole universe of science is built upon the world as directly experienced. . . .
>
> Science has not and never will have, by its nature, the same significance qua form of being as the world which we perceive, for the simple reason that it is a rationale or explanation of that world. . . . Scientific points of view, according to which my existence is a moment of the world's, are always both naive and at the same time dishonest, because they take for granted, without explicitly mentioning it, the other point of view, namely that of consciousness, through which from the outset a world forms itself round me and begins to exist for me. To return to the

---

an instrument with which to track down the building blocks of ideas by which consciousness functions. This method consists of imaginatively transforming a phenomenon in various ways so as to determine which aspects are essential to it and which not. We can imagine dogs with stripes and spots, with short ears and long ears, with pointed and flat noses—but never with wings or gills. In this way we can arrive at a general idea of "dog." This general idea can never be found in the world itself, but is a pure idea; the *Wesensschau* is a perception of the ideas used by thinking itself. The method of *Wesensschau* is part of the so-called eidetic reduction, a stage in Husserl's phenomenology in which a phenomenon in our consciousness is reduced to its "eidos," its form or idea. This eidos is a construction that must be presupposed in order to understand how human knowledge of reality is possible. The eidetic reduction belongs to an idealistic interpretation of phenomenology, in which Husserl had few followers.

> things themselves is to return to that world which precedes knowledge, of which knowledge always *speaks,* and in relation to which every scientific schematization is an abstract and derivative sign-language, as is geography in relation to the countryside in which we have learnt beforehand what a forest, a prairie or a river is. . . .
>
> The real has to be described, not constructed or formed. Which means that I cannot put perception into the same category as the syntheses represented by judgments, acts or predications. . . . The world is not an object such that I have in my possession the law of its making; it is the natural setting of, and field for, all my thoughts and all my explicit perceptions. . . . When I return to myself from an excursion into the realm of dogmatic common sense or of science, I find, not a source of intrinsic truth, but a subject destined to the world. (Merleau-Ponty 1962, viii–x)

At first glance these passages seem to address the same issues, while being affected by the same problems, as the more informal passage from Van den Berg. Merleau-Ponty introduces phenomenology as a method, as a way of describing the world that is an alternative, even *the* alternative, to the scientific method. From the phenomenological point of view, the scientific approach is a "rationale or explanation" of a more original world. Phenomenology returns to something more original—"To the things themselves!" (*Zu den Sachen selbst*), in Husserl's famous slogan. This is precisely the conception of phenomenology as method that is brought into question by the problems cited above. Certainly we must have "learnt beforehand what a forest, a prairie or a river is" before we can undertake scientific analysis and clarification of these things, but to say that phenomenology is in a position to describe these "things themselves" goes too far. Of necessity, any description of reality cannot avoid being a rationale, explanation, or constitution. That is not to say that the world is only a construction, just that we can never know the world as it is in itself, but only as we disclose it. An uninterpreted world, a world in itself, cannot be experienced; an untouched world cannot be lived in. Human beings never encounter a world in itself, only and always a world for them.

At the same time, however, it would be a mistake to dismiss phenomenology by virtue of this claim to make possible an originary encounter with the world. For something strange is at work in the above passages by Merleau-Ponty. While he claims again and again that phenomenology *describes* reality—and contrasts it with the sciences, which *analyze* it—he nowhere sets himself to producing such a description. What he actually does in the *Phenomenology of Perception* is to develop an analysis of the *relations* between human beings and their world, and he localizes this relation primarily in perception. Merleau-Ponty does not, then, describe the world, but rather the way in which human beings comport themselves to it. The "things themselves" that he addresses appear to be not the things of the world but rather the *relations between human beings and the world.* And in fact we find the same to be true of Husserl, the "founding father" of phenomenology, as well as of Heidegger, Husserl's least faithful but most influential student. Husserl tried, at least in his early work, to understand how human consciousness relates itself to the world. For him the "things themselves" were not objects in the world, but rather phenomena in consciousness, which form the way in which the world appears to us. And Heidegger, in *Being and Time,* did not describe the world itself, but rather inquired into the structures of the ways in which humans are engaged with the world in their actions and experiences.

It is, therefore, more in keeping with actual phenomenological practice to treat phenomenology as a philosophical movement whose principal task is to analyze the relation between human beings and their world rather than as a method of describing reality.[6] Thus I shall define "world" as "reality as disclosed by human beings"; the world-for-humans that arises when they act and experience it. Interpreted in this way, phenomenology sheds its claim to describe reality as it "authentically" is—and at the same time loses its vulnerability to contemporary philosophical criticism. Finally, this alternative interpretation of phenomenology opens up a new way to think about things.

Intentionality and Human-World Relations

How, then, does phenomenological analysis view the structure of the relations between human beings and their world? Although no single phenomenological method has been applied by all phenomenologists, a pattern can be discerned in the different approaches that phenomenologists take, a pattern that naturally does not do justice to the subtleties of the different philosophical positions, but that does indicate what they have in common.

The most important concept with which phenomenology works, and which should be preserved in a reformulation of phenomenology, is intentionality. In order to fully understand this concept, it is important to describe it in the context

6. Thus I join Ihde's pragmatically colored interpretation of phenomenology. See, for instance, Ihde (1979, 4–6; 1990, 3–25; 1993b, 5; 1998, 14–19).

in which it arose. In the epistemology of the nineteenth century, two movements developed—realism and idealism—with different perspectives on the relation between knowing subject and known reality.[7] Idealism awarded primacy to consciousness: all the knowledge that we have of reality was viewed as a product of consciousness. There can only be a reality when it is present in consciousness; reality appears as consciousness determines it. Realism, by contrast, assigns primacy to reality: all knowledge that we have of reality is a mirroring of the world itself. Consciousness then has genuine access to the world itself.

Phenomenology arose as an attempt to overcome the tension between idealism and realism. In this attempt, the concept of intentionality played the leading role. Husserl asked himself what is really given to human beings when they address themselves to the world. In answering this question, he found first of all that he had to suspend the "natural" attitude in which human beings assume that what is given to them corresponds to a world outside of them, or to an order fully articulated by reason. All presuppositions with respect to what is given must be put between brackets. This method of "putting things between brackets" Husserl called the *epoché*, or phenomenological reduction.

What remains left over of what is given to human beings, when the existence of a world outside of consciousness is put between brackets? First of all, it can no longer be characterized as a representation of a world, for the existence of a world can no longer be taken for granted. What remains left over are appearances, "phenomena"—whence the name phenomenology. But if this is the case, something else is given at the same time, namely, consciousness itself as the place where phenomena appear. But what more can profitably be said about this, at least without smuggling in new presuppositions? The following: that consciousness is directed to the phenomena that announce themselves in it. Human consciousness never exists in itself, but only as consciousness-of-something. It never exists as something isolated, but is always directed toward phenomena. This other-directedness is what Husserl calls intentionality. Intentionality is an essential characteristic of consciousness; a nonintentional consciousness is thus a contradiction in terms.

Phenomenology, however, did not remain a philosophy of consciousness. Husserl's followers, and even the later Husserl himself, came to believe that phenomenology needed to be more fully extended and worked out than a philosophy of consciousness.[8] Consciousness, consisting of knowledge about the

7. See Ihde (1976, 35–36).
8. See Ihde (1983, 122–35, 141–49).

world, came to be viewed as only one aspect of the relation between human beings and their world, and not necessarily the most relevant. Moreover, the world cannot be treated as an assemblage of objects for knowledge, but must be viewed as something in which human beings live: a lifeworld. Husserl's philosophy of consciousness broadened out into an analysis of the relation between humans and their world in the largest sense. In place of consciousness, for instance, Heidegger and Merleau-Ponty spoke about "being-in-the-world." Heidegger characterizes the intentional directedness to the world as having the structure of "care" (*Sorge*)—shaping one's own existence in the careful dealings with everyday things—while Merleau-Ponty views perception as a form of "being destined to the world." Husserl's followers refuse to restrict themselves to thinking only about knowledge, for this is only one of the forms of contact between human beings and world.

Phenomenology thus overcomes the dichotomy between subject and object, humans and world, by replacing it with a mutual interrelation. Human beings are unthinkable apart from a relation to the world, which they continually experience and in which they realize their own existence. This interrelation is not a fact that could have been otherwise. That was the point Merleau-Ponty was making in the above passage when he states that human beings "cannot put perception into the same category as the syntheses represented by judgments, acts or predications" and speaks of "a subject destined to the world." The focus on alienation so characteristic of classical phenomenology is absent from such phrases, and phenomenology is regarded as the analysis of the relation between human beings and their world. In order for a subject to render a judgment about reality, according to Merleau-Ponty, it must already be alongside and engaged with reality—which involves much more than judging, since it is the field in which judgments can take place. Human beings are continually engaged with their world, and this engagement precedes any judgment they may have of it. Put another way, it is impossible to speak about the world in the absence of human involvement with it. Reality-in-itself is unknowable, for as soon as we experience or encounter it, it becomes reality-for-us: a world. There exists neither human beings in themselves nor world-in-itself.

Phenomenology developed in this way not only by weakening its ties to the philosophy of consciousness, but also by establishing connections with existential philosophy. As mentioned in Chapter 1, existential philosophy, initiated by Kierkegaard, also consisted of an attempt to elucidate the relation between human beings and their world. It directed its attention not so much to the experiential aspects of this relation, but rather to the way in which humans realized their existence. One of its central insights was that human beings do not simply "exist" but have a relation to their own existence. Humans know that they exist and that they themselves need to shape their own existence. They can only do so in a world. The human way of existing is as "being there"; this existence always takes place somewhere. Existential philosophy, too, conceives human beings, therefore, via being-in-the-world, though in its efforts to elucidate being-in-the-world it emphasizes not the human experience of the world but rather the realization of human existence in it—human praxis or action.

The alliance between phenomenology and existential philosophy proved so fruitful that two perspectives on the relation between humans and world have crystallized out of it, one that approaches this relation from the perspective or "pole" of the world, and the other that takes as its point of departure the human "pole." The first analyzes the human-world relation in terms of the way in which the world can present itself to human beings and become meaningful; the second looks at the way in which humans are able to realize themselves in the world. The first perspective can be called hermeneutic phenomenological, inasmuch as it concerns interpretation and meaning—put most broadly, world-disclosure—and hermeneutics is the classical philosophical discipline that concerns itself with the disclosure of meaning. The second perspective, which concerns the way in which human existence takes shape, can be called existential-phenomenological.

Each of these two perspectives generates different philosophical questions about technology. In the hermeneutic perspective, the key question is the role technology plays in the way in which the world presents itself to human beings; in the existential perspective technology is described principally in terms of the role it plays in the way in which human existence takes shape. In my analysis of classical philosophy of technology in the previous two chapters I used the difference between these two phenomenological tasks as an implicit starting point, for Jaspers and Heidegger each occupy a different pole in the classical phenomenological approach to technology, with Jaspers representing the existential and Heidegger the hermeneutic pole.

TOWARD A POSTPHENOMENOLOGY OF THINGS

The above reinterpretation of phenomenology as the analysis of human-world relations makes it possible to overcome the dichotomy between idealism and realism in a more radical way than did classical phenomenology. While the latter bridged the gap between subject and object by stressing that, in fact, these two are always already intertwined thanks to the intentional engagement of human beings and world, a new interpretation of phenomenology can take this a step further by emphasizing that subject and object *constitute* each other. Not only are they intertwined, but they coshape one another. Human beings can only experience reality by relating to it, which does not involve any reality-in-itself but rather reality-for-them. As consciousness (perception, experience) can only exist as consciousness of something, reality is always reality for someone; in their engagement with reality, human beings always disclose it in a specific way. At the same time, humans themselves are constituted in this relation. The environment with which they are involved always codetermines in which ways they can be present to the world and each other. In the encounter between humans and world, each manifests itself in a particular way. In the mutual relation of humans and world there arises, therefore, a specific "objectivity" of world and a specific "subjectivity" of human beings.

Neither of these two poles can be absolutized. Human beings can not arbitrarily disclose any world, for there is always "something" that is disclosed—even if this "something" is inaccessible, just as was the case with Heidegger's dimension of "concealment," as discussed in the previous chapter. Were that not so, one could not speak of a relation between human beings and world, for the world would be a mere product of human beings. But neither are human beings arbitrarily constituted in this relation, for if "no one" manifests herself or himself in this relation it would be impossible to speak of a relation either—even if that "someone" cannot be present "in himself" or "in herself" but only in relation to a world. The fact that humans are what they are on the basis of their relation to the world does not imply that they are entirely determined by it.

This more radical phenomenological perspective, in which subject and object are not merely intertwined with each other but constitute each other, does justice to the contextualism of contemporary philosophy as it is expressed in the linguistic turn, in postmodernism, and also, for instance, in Latour's actor-network theory. I shall call this reinterpretation of phenomenology "postphenomenology." Ihde uses this term for his praxis-perception model of phenomenology, which revolves around the analysis of the perceptual aspects of the relation between human beings and their world.[9] In the introduction to *Postphenomenology,* he says that his philosophical orientation includes a strong sense of "proliferating

9. See Ihde (1993b, 3, 7; 1990, 30).

pluralism" and of the loss of centers and foundations (Ihde 1993b, 1), but he does not then go about showing what a reformulated phenomenology might look like under those conditions. This is the aim of the more radical interpretation of phenomenology that I am proposing, in which subject and object, or human beings and world, constitute each other. This interpretation can be called "postphenomenological" in that it overcomes both the essentialism and the fascination with alienation that characterized classical phenomenology.

Postphenomenology can be viewed as an offshoot of phenomenology that is motivated by the postmodern aversion to context-independent truths and the desire to overcome the radical separation of subject and object, but that does not result in relativism. From the postphenomenological perspective, reality cannot be entirely reduced to interpretations, language games, or contexts. To do so would amount to affirming the dichotomy between subject and object, with the weight merely being shoved to the side of the subject. Reality arises in relations, as do the human beings who encounter it. Only in this sense is postphenomenology a relativistic philosophy—it finds its foundation in relations.

Technological Intentionality

Postphenomenology offers a suitable framework for formulating a philosophy of technological artifacts that can resist the "Orphic temptation" to which the classical phenomenological philosophy of technology fell victim. Its perspective on artifacts, however, also needs to avoid the contrary of transcendentalism, namely, realism. For now that it is evident how problematic was the ambition of classical phenomenology to describe "the world itself," and now that it is clear that subjectivity and objectivity are constituted in the relation between human beings and their world, a turn "to the things themselves" runs the risk of landing the philosophy of technology back where it started. The ambition to think from the "things themselves" suggests the existence of an unmediated access to them.

But this suggestion is false. The facts that technological artifacts can be conceived as constructions, always exist in a context, and are interpreted by human beings in terms of their specific frameworks of reference do not erase the fact that systematic reflection can be undertaken of the role that these contextual and interpreted constructions play concretely in the experience and behavior of human beings. That "the things themselves" are accessible only in mediated ways does not interfere with our ability to say something about the roles that they play, thanks to their mediated identities, in their environment. And it is precisely the postphenomenological perspective that offers a new way of so doing.

In order to articulate the contours of a postphenomenology of things one can begin with the early work of Heidegger. As I have already shown, this work conducts an analysis of technology that stands in sharp contrast to his later philosophy of technology. In *Being and Time* Heidegger saw the relations between human beings and equipment as occupying center stage—or rather, he saw the role of tools and equipment as occupying center stage in the relation between human beings and their world (Heidegger 1996, section 16). Heidegger showed that tools and equipment give shape to the encounter between humans and their world. Things make daily practices possible while withdrawing from the explicit field of attention. Only when human beings occupy themselves not with their tools proper, but rather with what they set themselves to do with the help of these tools, are these tools present *as tools.* The tools are then, in Heidegger's words, "ready-to-hand."

This concept of readiness-to-hand directs our attention to the way in which objects are present in the relation between human beings and their world, and brings such things into precisely the domain that phenomenology investigates. The crucial question now concerns the various ways in which things, on the basis of their readiness-to-hand, play a role in the human-world relation. For such things shape this relation from their withdrawn or ingrown position, as has been shown by the examples already given. A train coshapes the way in which a landscape is present to human beings, a telephone coshapes the way human beings relate to each other. Things, therefore, are not neutral "intermediaries" between humans and world, but *mediators:* they actively mediate this relation.

Ihde has, from a phenomenological perspective, characterized this mediating role of artifacts in terms of what he calls *technological intentionality* (Ihde 1990, 141). By this he means that technologies—like consciousness for Husserl—have a certain directionality, an inclination or trajectory that shapes the ways in which they are used.[10] As an example, Ihde mentions the difference in writing style that arises when one writes with a fountain pen, typewriter, or word processor. One writes slowly with a fountain pen, with the result that it allows one to think over the sentence several times while composing it. The compositional speed is much faster with a typewriter, which tends to promote a style much closer to that of

10. For the sake of completeness, it should be noted that Ihde uses the term "technological intentionality" in another sense as well. By "technological" or "instrumental intentionality" he indicates the directedness of technologies toward specific aspects of reality. A cassette recorder, for instance, possesses a specific intentionality with respect to sound, which strongly differs from human intentionality, since it registers not only foreground but also background sound. Ihde also indicates this form of intentionality as a "technological telos." See Ihde (1979, 77–78; 1983, 56; 1990, 102–3).

spoken language. And a word processor, in contrast to pen and typewriter, vastly expands the ability to compose a text; for instance, sentences can be moved around and footnotes inserted at will. These writing technologies are therefore not neutral means, but rather play an active role in the relation between author and text. They have an intentionality, a trajectory that promotes a specific kind of use (140–43). They do not have a determining influence, for one can indeed write a slowly composed and carefully thought out text on a word processor, and write conversationally with a pen. But the technologies in question *promote* or *evoke* a distinct way of writing. Technologies, as it were, contain an "implicit user's manual" (Procee 1997, 159). A constructivist perspective on technology, as I shall elaborate further in Chapter 5, refers to this phenomenon as the "script" of technologies.

In the case of the fountain pen, this intentionality or innate trajectory became explicitly visible at the time of the introduction of the ballpoint pen. As the historian of technology Henri Baudet has pointed out, loud protests were made against the ballpoint pen when it first appeared. It was charged with having a negative influence on children's hand position and writing, and therefore on the quality of their work, the "neatness and care of their straight lines." Ballpoint pens were therefore viewed as "undermining instructional and pedagogical traditions." The classical way of writing with a fountain pen "represented a general social discipline," and this discipline was suddenly shattered by a faddish disposable product (Baudet 1986, 9–13).

Another example of the intentionality or trajectory of things is provided by an episode that happened in 1996 in the Romanian city of Cluj.[11] The mayor of this city proposed to shorten the shafts of the rakes used by the employees of the public gardens. These rakes, according to him, made possible an undesirable practice, allowing the employees to lean on them excessively. By shortening their shafts, the mayor thought, he could discourage laziness and encourage harder work. If the rake were merely a neutral means for the end of raking, this intervention would not have been necessary. Action had to be taken because the rake, en passant, made possible an entirely different practice, one that was not anticipated but that arose only in the practice of raking. The rake *mediates* the relation between the workers and the public gardens; it is not merely a means but plays an active role in the way this relation takes shape.

Ihde, to be sure, is not the only one to argue for the active role that things can play in their contexts. In the above brief sketch of Latour's actor-network theory

11. According to an article ("Shorter Shafts Combat Laziness") in the Dutch newspaper *Algemeen Dagblad* of 23 March 1996.

it was clear that for him things are active and can play full-fledged roles as "actors." The first philosopher of technology to devote extensive attention to the active role of artifacts was Langdon Winner. In his essay "Do Artifacts Have Politics?" (Winner 1986a), Winner described what has become a famous illustration of this principle, concerning the low-hanging overpasses on Long Island in New York.[12] These overpasses, designed by regional planner Robert Moses, were deliberately built low to prevent busses from using the roads and allowing only automobiles to pass underneath. The roads along which these overpasses were built lead to Long Island's beaches, meaning that these were now accessible only by car. At the time these bridges were built, this meant that racial minorities and the poor, who could not afford cars and generally relied on public transportation, were effectively prevented from reaching the beaches. Winner characterizes the role played here by the overpasses as "the politics of artifacts."[13]

The postphenomenological perspective described above allows a more radical extension of Ihde's concept of "technological intentionality." The "intentionality of artifacts" consists of the fact that they mediate the intentional relation between humans and world in which each is constituted. When human beings use an object, there arises a "technologically mediated intentionality," a relation between human beings and world mediated by a technological artifact.

Two different meanings of "intentionality" are therefore intertwined here, a first referring (in Ihde's sense of "technological intentionality") to the "intentions" of the technology itself, the second (in the more general phenomenological sense of "technologically mediated intentionality") to the relations between human beings and world that are mediated by the technology.[14] Both meanings are relevant for a phenomenological understanding of the role of technologies in human-world relations. When technologies mediate the intentional relation between humans and world, this always means from a phenomenological perspective that they codetermine how subjectivity and objectivity are constituted. Their "intentionalities," in Ihde's sense, consist of the fact that they coshape the contact between human beings and their world; they determine how human beings can be present in the world, and the world to them.

12. See Winner (1986a) and Smits (2001).

13. Winner's example has recently been challenged. It turns out, for example, that the overpasses in question probably have never been an obstacle for buses, as can be shown with the help of timetables. See Woolgar and Cooper (1999).

14. In fact, three meanings of intentionality are intertwined here, if one includes the additional meaning referred to in note 10 above.

Multistability

There is, however, one pitfall that needs to be avoided in this analysis of the ability of artifacts to coshape the relation between human beings and world: this ability must not be conceived as an intrinsic property of the artifact itself. The effect of this misconception would be to smuggle back in again via the back door the old subject-object dichotomy—which it was precisely the triumph of phenomenology to have overcome. It would give rise to a kind of realism in which properties would be assigned to objects independently of the subjects for whom these objects exist. Winner's example of Robert Moses's overpasses makes clear, however, the shortcomings of such an approach. For the politics of these overpasses has considerably diminished with time. In a role reversal, the poor, too, now own automobiles, while many wealthy families take their vacations in campers big enough to be barred from traveling on the parkways in question (Achterhuis 1998, 386).

The thought that technological artifacts possess intrinsic properties and can themselves influence the relation between human beings and world supposes that technology can be spoken about independently of the humans that engage with it. But from the phenomenological perspective this is untenable. Artifacts can only be understood in terms of the relation that human beings have to them. Here one can make the same phenomenological move that others in that tradition make with respect to "consciousness" and "perception." Just as "perception-in-itself" and "consciousness-in-itself" do not exist, neither does "technology-in-itself." Just as perception can be understood intentionally only as perception-of, and consciousness only as consciousness-of, so technology can only be understood as technology-in-order-to. The "in order to" indicates that technologies always and only function in concrete, practical contexts and cannot be technologies apart from such contexts. In Ihde's words, "Were technologies merely objects totally divorced from human praxis, they would be so much 'junk' lying about. Once taken into praxis one can speak not of technologies 'in themselves,' but as the active relational pair, human-technology" (Ihde 1993b, 34).

The insight that technologies cannot be separated from their use contexts implies that they have no "essence"; they are what they are only in their use. A technology can receive an identity only within a concrete context of use, and this identity is determined not only by the technology in question but also by the way in which it becomes interpreted—as shown by Robert Moses's overpasses. Another example illustrating the context dependence of technologies is to be found in the early development of the typewriter, driven as it was by the desire to

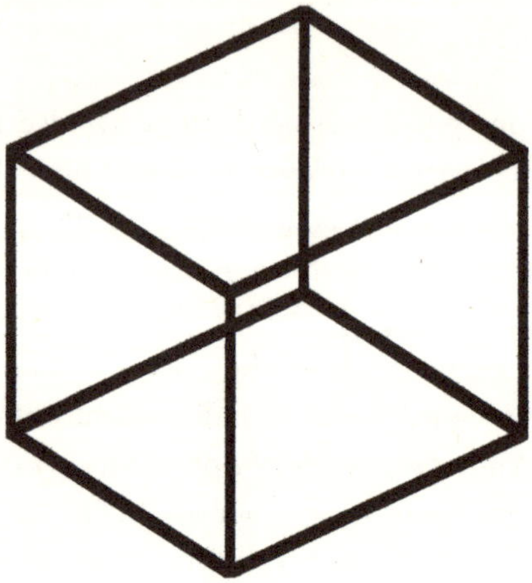

Fig. 1 The Necker cube

design equipment for the blind and partially sighted. But it quickly took on another identity, as a writing technology that is useful for nearly everyone (Ihde 1993a, 116).

Ihde calls such context dependence "multistability," and to clarify what he means he makes use of a perceptual example, the so-called Necker cube (figure 1). When we look at this figure, we can see more than one thing. Sometimes we see a three-dimensional cube with the bottom surface and two side surfaces turned toward us. If we try, we can switch between the two manifestations of the cube. We can also interpret the figure two-dimensionally and see it as an insect with six legs sitting in a six-sided cell of its web. Ihde uses this example to illustrate that different ways of seeing produce different figures. The figure allows multiple interpretations. What it "really" is remains undetermined. It is many things at once; it is "stable" in multiple ways. Something similar, according to Ihde, is at work in connection with technology. Just as the Necker cube has no "essence," neither do technologies. They are only technologies in their concrete uses, and this means that one and the same artifact can have different identities in different use contexts.

Two Dimensions

A postphenomenological "turn toward things" in the philosophy of technology, as indicated above, needs to consist of the analysis of the mediating role of technological artifacts in the relation between human beings and reality. Such an analysis can be carried out in the phenomenological territory just described—hermeneutical and existential. In both dimensions of the intentional relation between human beings and world artifacts play a role, to be further specified in the chapters to come.

In hermeneutical terms, things can mediate the ways in which human beings have access to their world by the roles that such things play in human experience. Questions such as the following arise: In what way do telescopes and electron microscopes, automobiles and airplanes shape our access to the world? In what way are others present to us when we contact them via telephone or email? An analysis of the technological mediation of our experience produces a new interpretation of hermeneutics. In place of the traditional emphasis on language and text, in this "material hermeneutics" things take center stage. In the next chapter I shall use the work of Ihde to elaborate this postphenomenological perspective on technology.

In existential terms, things mediate human existence. Here a different set of questions arises: How does the television set affect the way we divide up our day? What implications do automobiles and airplanes have for the way in which we organize our social relations? In seeking the answers to such questions, the existential-phenomenological perspective can acquire a more material interpretation. In this interpretation, concepts such as authenticity become less central, and more attention is paid to the way in which the material environment of human beings shapes the way in which they realize their existence. I shall work out this existential dimension of the acts of artifacts in Chapters 5 and 6, in a discussion of the work, respectively, of Bruno Latour and Albert Borgmann. Their positions contrast sharply with each other, but each highlights important aspects of the role of things in human acts and existence, which can be further elaborated from the postphenomenological perspective.

4

A Material Hermeneutic

INTRODUCTION

By now it should be clear that a philosophy of technological artifacts will arise from an inquiry into the role that artifacts play in the mutual relations between human beings and world. Of the two postphenomenological perspectives sketched out in the previous chapter, I shall take up and elaborate the hermeneutical one in this chapter. More specifically, I shall elaborate a perspective on technology in which the classic hermeneutical orientation toward texts is broadened to encompass things and artifacts. Its central question is, "What role do technological artifacts play in the manner in which human beings interpret reality?" The work of the American philosopher of technology Don Ihde can serve as a point of departure for responding to this question.[1]

Ihde was one of the first to have brought phenomenology explicitly in contact with the philosophy of technology, by applying central phenomenological concepts and methods developed by Husserl, Heidegger, and Merleau-Ponty. After a short introduction to his specific elaboration of

1. For an earlier overview of Ihde's work, see Verbeek (2001, 119–46), which has been thoroughly supplemented and revised here.

phenomenology and hermeneutics, I shall address how he developed a philosophy of technology in which technology is approached in terms of artifacts. Central to Ihde's approach is an analysis of the different relations human beings can have with different kinds of technologies. On the basis of Ihde's analysis it becomes possible to investigate the role of artifacts in the hermeneutical dimension of the relation between human beings and their world. Technologies help to shape the way in which reality is present to human beings; not only how they perceive the world, but also the frameworks in which they interpret it.

RELATIONS BETWEEN HUMAN BEINGS AND ARTIFACTS

Ihde's conception of phenomenology approaches the relation between human beings and their world in terms of experience, since this is the place where the mutual relation between humans and world can be localized. Experience is, as it were, the interweaving of both: in experience, human beings and world—or subject and object, for that matter—are not separated but are always intertwined. Only afterward, when an experience is described and not enacted, does it make sense to separate out a perceiver and a perceived or a subject and an object. In the perceiving itself that cannot be done, for to perceive is to perceive *the world.*

In his analysis, Ihde directs himself first of all to the *structure* of experience (Ihde 1976, 2; 1990, 21–23). He calls this approach *relativistic* (1998, 46; 1990, 23–25)—not in the sense of an epistemological relativism, but rather in the more literal sense of an *analysis of relations.* "A phenomenological account . . . always takes as its primitive the relationality of the human experiencer to the field of experience. In this sense, it is rigorously relativistic. The relationality of human-world relationships is claimed by phenomenologists to be an ontological feature of all knowledge, all experience" (Ihde 1990, 25). Experience requires the existence of a relation between the experiencer and what is experienced; it supposes, that is, "intentionality." Experience is always experience-of-something, as described in the previous chapter. The experiencer and the experienced are inextricably bound up with each other.

Ihde distinguishes two dimensions of experience. The first is the bodily dimension of sensory perception, which Ihde calls *microperception.* The second consists of the frameworks within which sensory perception becomes meaningful. Human experiences can be conceived as "interpreted perceptions," in which the interpretations are always informed by the cultural context in which they take place. This contextual dimension of experience Ihde calls *macroperception:*

> What is usually taken as sensory perception (what is immediate and focused bodily in actual seeing, hearing, etc.), I shall call microperception. But there is also what might be called a cultural, or hermeneutic, perception, which I shall call macroperception. Both belong equally to the lifeworld. And both dimensions of perception are closely linked and intertwined. There is no microperception (sensory-bodily) without its location within a field of macroperception and no macroperception without its microperceptual foci. (29)

While it is true that microperception and macroperception can be distinguished from each other, they cannot be separated. A bodily perception can no more exist without being interpreted than an interpretation can exist without something to be interpreted. The twofold meaning of the word "perception" to which Ihde points is illustratively present in the verb "to see," which we can use to describe a bodily-sensory perception ("I see a tree") and to characterize an interpretation of the world ("Since that talk I see things completely differently").

Ihde pursues his analysis of the role of technology in the interrelation of human beings and world by inquiring into the forms of these interrelations when technological artifacts are involved. To that end he distinguishes three different ways in which human beings can relate to technological artifacts. In the first, our perception is mediated by a technological artifact. In such a *relation of mediation* we are not directly related to the world but only are so via an artifact—as for instance whenever we wear glasses or watch television. A second kind of relation, which Ihde calls an *alterity relation,* is a relation not via an artifact to the world but to an artifact itself, as for instance when operating a machine. The third kind of human-technology relation Ihde calls a *background relation,* in which technological artifacts shape our relation to reality but do so by remaining in the background, as do thermostats that automatically switch the heat on and off without our intervention or even awareness.

Relations of Mediation

Technologies play an important role in our daily lives by mediating our experience. We read the temperature via thermometers, we remember events via photographs, we speak with each other via telephones, and we keep abreast of current events via television. In all of these cases we are not *directly* in bodily-sensory experience present to the world but are so *via* technological artifacts. How is this relation between human beings and artifacts to be understood? Ihde begins to work out his answer to this question with the aid of Heidegger's analysis of tools and Merleau-Ponty's analysis of the role of "embodiment" in perception.

Heidegger's analysis of equipment, discussed in Chapter 3, arrives at the conclusion that a tool is "something in order to"; it is serviceable, helpful, useful (Heidegger 1996, 64). A hammer is most present to us as a hammer when we hammer with it rather than use it, for instance, as a paperweight. Tools have a way of being that Heidegger calls "handiness" or "readiness-to-hand" (*Zuhandenheit*). It is characteristic of something ready-to-hand that it withdraws itself from our attention in order to be used. Someone who is hammering is not concerned with the hammer but rather with what is being done, or made, with the hammer.

Ihde finds in Heidegger's analysis of the ways in which tools are present for human beings three elements of special significance. First, Heidegger shows that each tool, each piece of equipment is related to a context. In itself it is nothing; as a piece of equipment it is part of a meaningful whole. "This field within which a tool is what it can be is a complex one filled with 'involvements' or cross-relations" (Ihde 1990, 32). Second, it is clear from Heidegger's analysis that equipment has an "instrumental intentionality"; a tool is "something in order to," and in that "in order to" there is always a reference of the tool to a use context, to whatever can be done with it (32). By the concept of "intentionality" here Ihde means the directionality of the instrument and not the intentional relation between human beings and world, to which the phenomenological term usually refers.[2] The third element that Ihde finds of special significance is that the tool or piece of equipment, in use, becomes the *means* rather than the *object* of our experience.

One might well question this third element, given that in his analysis of equipment Heidegger is not referring to its role in *experience* but rather in *praxis*—Heidegger investigates what it practically makes possible, withdrawing in the process. He refers to the hammer not as a means to *experience* the nail but rather to drive it in the wall. The idea that artifacts also mediate *experience* is more fully worked out by Merleau-Ponty, another thinker from whom Ihde draws much. Ihde finds in Merleau-Ponty's work two examples that he considers especially illuminating concerning how human beings are related "through objects" to the world: the examples of a woman with a feather in her hat and the example of a blind man with a cane. Merleau-Ponty uses these examples to show that human beings can use artifacts to stretch the spatiality of their bodies.

A woman with a feather in her hat can extend her area of sensitivity to the world to the point where she can keep a safe distance between the feather and

2. See the end of Chapter 3 for a closer examination of this double meaning of "intentionality."

objects that might damage the feather, stooping instinctively when necessary; "she feels where the feather is just as we feel where our hand is" (Merleau-Ponty 1962, 143). The image of the blind man's cane carries this a step further, making it clear that human beings can not only *extend* the spatiality of their lived bodies with the aid of artifacts but *perceive* with them as well. With his hand, a blind man feels not so much the stick as the street and the objects in the way through the stick.

These images provided by Heidegger and Merleau-Ponty are in effect complementary. While Heidegger analyzes the ways in which artifacts *are present* to human beings, "withdrawing" from their experience, Merleau-Ponty analyzes the *relations to the world* that can arise on the basis of this presence. These analyses, taken together, point to a structure of perception that can be described in terms of *mediation*. The intentional relation between human beings and world is thus, as it were, extended or stretched out through artifacts. Ihde schematizes the difference between unmediated and mediated perception as follows:

unmediated perception:	I–world
mediated perception:	I–technology–world

It must be emphasized that by "unmediated" Ihde means unmediated by artifacts. All perceptions are in a certain sense mediated, because human beings have access to the world only via interpretation. Ihde is not concerned here with mediation of this type; when he speaks of "naked perception," he means not some preinterpretive access to reality but a perception that takes place without the intervention of an artifact on the microperceptual level.

In analyzing a number of examples of mediated perception, Ihde comes to the conclusion that there are two basic sets of relations with artifacts in which they mediate people's relations with their world. The first involves what he calls *embodiment relations*. In these relations, human beings take technological artifacts into their experiencing, and thereby broaden the area of sensitivity of their bodies to the world. An example of the embodiment relation is the wearing of eyeglasses. When I wear eyeglasses, I do not look *at* them but *through* them at the world. I take the pair of glasses into myself; it withdraws from my perceiving. But embodiment relations are not restricted to the visual. A dentist, for example, who uses a dental probe to feel out cavities in my teeth is using the probe to extend the sensitivity of touch, feeling cavities via the probe. Ihde schematizes embodied relations as follows:

embodiment relations: (I–technology) ⟶ world

The most important characteristic of embodied technologies is that they possess a certain transparency. They call attention not to themselves, but to (aspects of) the world given through them. In order for this transparency to occur, however, several conditions must be met: (1) The artifact must be technically serviceable; that is, its physical characteristics must allow it to be embodied. A pair of glasses made with opaque lenses cannot serve embodied perception; (2) a certain skill or technique is required to perceive through the artifact; those not trained in dentistry cannot use dental probes to detect tooth decay; (3) the artifact should aim at making mediated perception take place according to a measure comparable to unmediated perception; a telescope ordinarily delivers a picture of a planet with roughly the same size as the picture of a red blood corpuscle delivered by a microscope—"the image size of galaxy or amoeba is the same" (Ihde 1990, 79).

The second set of mediated relations with artifacts that Ihde considers consists of *hermeneutic relations.* In hermeneutic relations, too, we are involved with the world *via* an artifact, but the artifact is not transparent. The artifact does not withdraw from our relation to the world but provides a *representation* of the world, which requires interpretation in order to impart something to us about it. Because this relation involves interpretation (the artifact must be "read"), Ihde calls it *hermeneutic.* Hermeneutics, after all, is the analysis of interpretations. In hermeneutic relations the world is not perceived *through* the artifact but *by means of* it. Ihde schematizes hermeneutic relations as follows:

hermeneutic relations: I ⟶ (technology–world)

An example of a hermeneutic relation with an artifact is the use of a thermometer. When we read a thermometer, we are not involved with the thermometer but with the world, of which the thermometer reveals one aspect, namely, its temperature. This revealing, however, does not have the character of a sensing of temperature but is rather a *representation* of it.

The technological mediation of perception, whether embodied or hermeneutic, has consequences for the ways in which human beings experience their world: artifacts *transform* experience. Later in this chapter I shall develop the hermeneutic implications of this insight.

Alterity Relations

The second human-technology relation described by Ihde is the alterity relation. In alterity relations humans are not related, as in mediating relations, via a technology to the world; rather, they are related to or with a technology. The role played by technologies in this set of relations can be characterized as that of a "quasi-other." This set of relations can be formalized as follows:

alterity relations: I ⟶ technology (–world)

Technology appears in alterity relations as quasi-other because while we may encounter technologies in ways in which they seem to behave as an "other," they can, of course, never be present as a true person. Humans often approach the technologies that they encounter in anthropomorphic ways: they project human properties onto artifacts ("intelligent computers"), or entertain certain feelings for them ("caring for" a piano). But a technology is never a genuine other. An automobile may be seen as something that I care for, but it is far less of an other even than a horse, which can also be used to travel but which does not always obey and can even be startled or rear if a rabbit happens to cross its path. The automobile is at best a quasi-other.

The reason that technologies in alterity relations are experienced as quasi-others is that on the one they hand possess a kind of independence and on the other hand they can give rise to an "interaction" between humans and technologies. Many toys, such as tops and music boxes, are fascinating precisely because of the apparent autonomy that they possess. Robots and automatons possess such autonomy, to the extent that one could truly speak of "interacting" with these technological "beings." Automatic train ticket machines, for example, not only take money and dispense tickets, but also give advice, provide route information, answer questions, and protest when something is done incorrectly.

Embodiment relations, hermeneutic relations, and alterity relations, according to Ihde, form parts of a continuum. On the one extreme of this continuum are embodiment relations, in which a technology has the role of a quasi-I. In embodiment relations technology always coincides, as it were, with myself. At the other extreme are alterity relations, in which a technology is present as a quasi-other, as indicated above. Between these two are hermeneutic relations, in which the technology mediates and is therefore not present "as itself," but at the same time draws attention to itself because it is not embodied but "read."

Background Relations

The final set of human-technology relations that Ihde identifies is formed by background relations. In contrast with the two kinds of relations already discussed, technological artifacts in background relations do not play a central role in our experience. In background relations, we are related neither explicitly to a technology nor via a technology to the world; instead, technologies shape the context of our experience in a way that is not consciously experienced. Schematically, these relations can be expressed as follows:

background relations: I (–technology/world)

Refrigerators and central heating systems are examples of technologies with which we can have a background relation. These technologies switch themselves on and off in the background of our experience; we notice scarcely if at all that the room temperature is almost always the same and that the refrigerator is on. Technologies such as these give rise to a background "field" in which we can have experiences without explicitly experiencing the technologies in question. They are present and absent at the same time: without us noticing them, they give form to our experience by shaping a context for it. As such, they can have many of the same transformational characteristics as the other technologies mentioned with which our involvements are much more explicit. Their "absent presence" is usually experienced only when they stop functioning—when a storm knocks out the electricity, for instance. In such cases the context shaped by the background technologies, which we otherwise take for granted, is suddenly not self-evident any more.

MEDIATION AND MEANING

Clarification of the different types of human-artifact relational structures lays the groundwork for a new approach to the hermeneutical perspective within (post)phenomenological philosophy of technology, for it exposes different roles that artifacts can have in the relation between humans and world. The principal hermeneutical question concerning technology is: what role do technologies play in the way in which human beings interpret reality—or, conversely, in the way in which reality comes to be meaningful for human beings?

The answer to this question can have two forms, depending on whether one is inquiring into *direct* or *indirect* ways of mediation in the origin of meaning. The direct way concerns the mediation of sensory perception; by shaping the way in which humans perceive reality in microperception, that is, artifacts help to determine the possible ways in which it can be interpreted. The indirect way concerns macroperception, or the technological mediation of frameworks of interpretation that coshape the relation of humans to their world. In what follows, I shall use Ihde's work to explore the role of artifacts in two such frameworks: the cultural framework in which human beings find themselves and the scientific framework in which technological instruments play an indispensable role.

Beyond Subject and Object

Of all the human-technology relations that Ihde analyzes, the relations of mediation (the embodiment relation and the hermeneutic relation) are especially relevant to investigating the hermeneutical implications of perceptual mediation. After all, these are the main relations from which technologies shape the way in which reality is present to human beings. But before proceeding any further, it is necessary to hone Ihde's analysis on one point. For the way in which he speaks about technological mediation seems at times to lapse once again back into the subject-object schema that it is precisely the phenomenological ambition to be overcome.

By saying that mediation is located "between" humans and world (as in the schema I–technology–world), Ihde seems to put subject and object over against one another, instead of starting from the idea that they mutually constitute each other. His analysis appears to suggest that he takes as a point of departure humans already given as such and a world already given as such, in between which one can find artifacts. Ihde does not address this problem in *Technology and the Lifeworld,* though it gnaws at the roots of his approach to the phenomenology of technology. Only later, in *Expanding Hermeneutics,* does he make clear—completely in line with the postphenomenological perspective—that subject and object are mutually interrelated,[3] but he does not connect this thought with his earlier analysis of human-technology relations.

The phenomenological insight that subject and object are mutually interwoven thus makes it necessary to supplement Ihde's analysis of technological mediation. It might be tempting to conceive mediation as a process in which a transformation occurs of the manner in which a subject (or human being) experiences an object (or world)—in other words, as a process between a fixed subject and a fixed object wherein only the manner in which the object is experienced by the subject is affected. But from a postphenomenological perspective mediation cuts deeper than that. The relation between subject and object always already precedes

3. In the chapter "Interpreting Hermeneutics," Ihde states that "Subject and object arise within what Husserl sometimes calls the *correlation-apriori* which began as intentionality" (Ihde 1998, 16).

the subject and the object themselves, which implies that the subject and the object are mutually constituted in their interrelation. In any relation between subject and object, both are brought into existence in a specific way, and both subjectivity and objectivity acquire specific shape.

When analyzing the mediating role of artifacts, therefore, this mediation cannot be regarded as a mediation "between" subject and object. Mediation consists in a mutual constitution of subject and object. This must be borne in mind when considering Ihde's discussion of the various relations between humans and artifacts. Mediation shapes the mutual relation in which both subject and object are concretely constituted. Someone who wears eyeglasses, for instance, is not the same without them. When I wear glasses, or some other equivalent like contacts, I am in the world differently than without them. Without glasses I cannot play the piano or drive a car, and I write rather poorly. My world and the way I am present in it is profoundly shaped by my glasses. Wheelchair users experience this mediation of subjectivity and objectivity still more strongly. Many of their everyday, worldly activities, from greeting another person to carrying on a conversation to partaking in meals, are shaped to a large extent by their wheelchairs. The same can be said for the way in which their world is present to them; many objects are present as obstacles, and wheelchair users experience the world from a sitting perspective, which determines, for instance, a specific kind of relation to conversational partners.

In none of these cases is it a matter of already given subjects and objects, between which artifacts are inserted. What humans are and what their world is receive their form by artifactual mediation. Mediation does not simply take place *between* a subject and an object, but rather *coshapes* subjectivity and objectivity. Formulations in terms of the "access to reality" offered by an artifact should be read as relating to the way in which an artifact makes possible the constitution of a world in the very process of perception. Humans and the world they experience are the *products* of technological mediation, and not just the poles between which the mediation plays itself out.

Transformations of Perception

But how does this work on a concrete level? What are the implications of technological mediations for our experience—for the way in which the human world acquires meaning? Ihde's analysis of the way in which technologies mediate human perception offers a point of departure for such questions. Mediation, for Ihde, is indissolubly linked with a *transformation* of perception. Naked perception and perception via artifacts are never completely identical. In this transformational character of technological mediation lies an important aspect of the non-neutrality of technology.

This transformation of perception has, according to Ihde, a definite structure involving amplification and reduction. Mediation always strengthens specific aspects of the reality perceived and weakens others. When we use a telephone, for instance, we are speaking to our interlocutor at a distance that previously would have been unthinkable. At the same time we are only audibly present to each other, not visually or in a tactile way. When we look through a telescope at the moon, we can distinguish details that otherwise would be invisible. At the same time, the moon is detached from the context of the night sky with its other heavenly bodies, which is where the naked eye comes across it; the moon's image is "enframed." Only when the telescope is put aside does that context return.

The transformation of perception, with its structure of amplification and reduction, appears in different gradations. When we compare mediated perception with perception with the naked eye, we can distinguish between transformations of low contrast and transformations of high contrast. The transformation that a pair of eyeglasses brings about, for instance, is a transformation of low contrast. The world that eyeglass wearers perceive strongly resembles the world that they saw before they needed glasses; the only difference is that the image is enframed. But the transformation effected in a spectrogram, for instance, involves a high contrast with respect to naked perception. A spectrogram is a visual deposition of the light given off by, say, a star, from which information can be derived about the star's chemical composition. The spectrogram's band of colored stripes is as removed as possible from the star that we see with the naked eye; nevertheless, it reveals important aspects of the star. A thermometer, too, provides a transformation of high contrast; in place of feeling warmth and cold we read a number to establish temperature.

Mediated Meaning

What are the hermeneutical implications of this transformation of perception? What have artifacts to do with the way in which humans and world are constituted? Ihde's work provides us with a good start toward answering these questions but ultimately leaves us stranded. When he analyzes the hermeneutical implications of perceptual mediation, he does so by investigating the implications for cultural and scientific frameworks. He shows, in other words, how transformations of microperception affect macroperceptual ways of seeing. But mediations also have hermeneutical implications for the microperceptual level. Artifacts help to shape human interpretations of reality not only because they play a role in interpretive frameworks, but also because of their role in sensory perception, which determines the very possibilities human beings have for interpreting reality.

Dutch psychiatrist and philosopher J. H. Van den Berg offers an initial, though rather problematic, entry into the hermeneutical aspects of microperceptual mediation. In his book *Things: Four Metabletic Reflections,* he devotes a passage to the telescope and the microscope wherein he tries to elucidate the way in which reality is present to us through such instruments. As noted in the previous chapter, Van den Berg adheres to the classical phenomenological view that science and technology involve an impoverishing of reality with respect to what it should "authentically" be. This diagnosis, which the postphenomenological perspective reveals to be untenable, is also present in Van den Berg's book. Nevertheless, his views on mediated perception can provide a heuristic tool for connecting Ihde's thoughts about the transformation of microperception with the question of the role that technology can have in the generation of meaning.

Van den Berg describes himself standing on a balcony using a pair of binoculars to peer at a wooden baluster that was part of the balcony and is now lying in the garden below. He asks himself, "What do the binoculars do to the observed object?" At first he points to the *reducing* aspects of the transformation the binoculars bring about. Through the binoculars he sees "a simmering, spherically limited, small piece of more or less meaningless grass, on which lies a slightly unfamiliar piece of wood with markings." In this perception, the context in which the baluster was present for the naked eye—"the garden in the summer"—had vanished. "My binoculars cannot encompass the garden; even the lawn alone is too large for that." Van den Berg finds the amplificatory aspects of the perceptual transformation to be a form of reduction. Not only do the binoculars lose "the whole" of the unmediated experience, he says, but the reality disclosed in the intensified perception is quite alien to everyday reality. Binoculars, according to him, lose the "dimension" of the thing. "To enlarge is to see outside the dimension which the things have." In unmediated experience "things are as they are; they have their own dimension. And if they didn't what could *dimension* then possibly be?" The binoculars strip the baluster "of its dimension, along with its own character. This, too, can be seen. The baluster which I see through the binoculars possesses an unreal character" (Van den Berg 1970, 9–10).

Van den Berg goes further, and after describing the wooden baluster as viewed through binoculars considers what it would be like to view it with a microscope. A microscope that enlarges five hundred times, he points, out, would make him see the piece of wood as a tree some five kilometers tall—not something one finds in our daily lifeworld. Human beings, he concludes, do not see a *tree* through a microscope; rather, they see *cells,* which tell them nothing about the tree as they experience it in their ordinary lives. We experience cells only when we look through the microscope, and they are not necessary to experience or understand a tree. "The microscope isolates the tree from [its] context, for it strips the tree of its dimension and, by the same token, deprives it of its meaning and destiny. Its cells in themselves have no meaning, but the tree does" (12–15).

According to Van den Berg, these cells are part of a secondary structure of reality, which, in its turn, does not need the tree as it is present in everyday life. Van den Berg contrasts this abstract-scientific reality with the "first" structure of everyday experience. It is fully possible to change over from the one kind of structure to the other: with the aid of a microscope, for instance, one can establish whether or not a tree is sick and consequently deal better with the tree in the everyday experience of it. But in that case "it would be as though I would use a necessary detour to reach that purpose, as though I would travel to my destination by way of a foreign country. The trip offers numerous advantages, but at the same time the danger that existing boundaries will be disregarded. One could easily be busy with the second structure while erroneously thinking that one is learning details about the first" (13–14).

Thus a microscope, according to Van den Berg, discloses reality in a detached context of meaning that is alien to the world in which human beings go about their daily lives. This brings us back to the problematic, classic phenomenological notion that science and technology deliver a reduced reality, and runs dead against the postphenomenological approach that I am developing, in which technology provides specific forms of *access to* reality. Van den Berg acknowledges that there is a direct connection between perceptual mediation and meaning—a connection that is underexposed in Ihde's work—but the way he does so is problematic. Van den Berg concludes too quickly that the technological mediation of reality necessarily entails a disclosure of reality in a secondary structure, which is a reduction of the everyday, primary or "first" structure.

Moreover, it is incorrect to describe the secondary structure as impoverished with respect to the primary. When the way in which technologies mediate perception is looked at more carefully, a more nuanced picture is required of the relationship between mediation and meaning. The perception achieved by a pair of eyeglasses, for instance, that effects a transformation of extremely low contrast, provides access to reality in practically the same domain and with the same possibilities of interpretation as a perception that is not mediated by technology. A pair of binoculars and a telescope do that to a lesser extent, in that some of "the whole" of experience available in unmediated perception must be given up. They transform the reality that they make visible into something that can only be visually interpreted, and remove the perceived out of the context in which it exists for the naked eye. At the same time these technologies open worlds that previously had been hidden. The transformation effected by a microscope is still greater because it makes perceptible a reality that deviates more strongly from our daily reality. A spectrogram, finally, reveals reality only in terms of specific scientific phenomena, which further restricts the number of possible interpretations. It reveals only one aspect of reality, and a scientific one at that: the chemical constituents of the phenomena perceived.

The mediation of perception therefore does not necessarily lead to a reduced picture of reality. Instead of "primary and secondary structures," it is better to follow Ihde in speaking in terms of "transformations of high and low contrast" when comparing mediated perception with naked perception. When doing so, embodiment relations and hermeneutic relations should be viewed as the extreme ends of a continuum. As we move on this continuum away from embodiment and more toward hermeneutic relations, the transformation that reality undergoes in the mediation is one of progressively higher contrast: the perception offered by the mediation deviates ever more sharply from unmediated perception. The reason for this is that what mediated perception can make visible is determined with ever more specificity as the mediation grows more hermeneutic in nature. A hermeneutic technology, after all, provides a representation of reality, which implies that the design of such a technology predetermines which aspect of reality is to be made perceptible by it and in which ways.

This, however, does not necessarily mean a reduction. Reduction is the flip side of amplification, since there always occurs a strengthening of aspects of reality as well. Technologies can even open up new ways for reality to manifest itself. Ihde points to infrared photography as an example. In this form of photography we lose, to be sure, the nonvisual aspects of the photographic object and the depth of the unmediated image, but at the same time it makes perceivable things that remain invisible to the naked eye (Ihde 1991, 73–74). It is easier to tell, for example, whether trees are diseased on the basis of infrared photographs of them than through inspection with the naked eye.

The insight that technologies can play a mediating role in our experience, in which certain aspects of the world are strengthened and others weakened, points to the need to fine-tune the classical-phenomenological thought that technology consists of a specific, and reduced, interpretation of the world—a thought shared not only by Van den Berg but also by Heidegger, for whom technology is characterized by an interpretation of the world as "standing-reserve," or *Bestand,* a storehouse of raw materials that lies ready for human manipulation. This interpretation would allow the world to appear only in a very limited respect: under the guise of control and domination. But when we consider technology in terms of concrete, mediating artifacts, as does Ihde, it becomes clear that our dealings with these artifacts do not require us to have such a "controlling" interpretation of the world. A tree is not forced to show itself as firewood or as potential furniture material when viewed through a pair of eyeglasses; rather, the pair of eyeglasses opens up to its wearer the same domain of possibilities of interpretation as are available to the person who doesn't wear eyeglasses. And certain technologies can even allow the world to manifest itself in new ways.

Technologies are therefore more ambivalent than alienating with respect to the interpretations of the world with which they are linked. When they mediate our experience, they have as much a reductive as a strengthening impact on our experience. The more it is possible to embody a technology, the less it predetermines in which ways the world can manifest itself through it. But even if a technology does determine these ways of manifestation, this does not necessarily entail a reduction. Technologies also make possible new modes of access to reality that would be impossible without mediation. Technologies thus constitute a new reality, a new "objectivity."

ARTIFACTS, CULTURE, AND SCIENCE

The hermeneutic dimension of mediation by artifacts should not only be localized in microperception, but also is found in the macroperceptual frameworks in which human beings find themselves. The ways in which reality comes to be meaningful for human beings depend not only on their sensory relations with it but also on the contexts in which meaning arises. One finds in Ihde's work two macroperceptual contexts in which forms of technological mediation play an especially important role: the cultural context in which everyday human life plays itself out and the scientific frameworks of interpretation that, to be sure, play an increasing role in everyday life.

Cultural Mediation

Ihde's approach offers an alternative to the way the classical philosophy of technology has formulated the relation between technology and culture. For to approach technology in terms of mediation reworks the traditional dyad of instrumentalism and substantivism.[4] Instrumentalists see technology as a mere tool, as the means to accomplish certain tasks. Within this approach technology appears as something neutral; if technology is viewed as a means to an end, this implies that it is not to be judged in itself but only in its use. Substantivists, on the other hand, think that instrumentalists overlook the fact that technology is not at all neutral. According to this school of thought, technology must be understood as an independent power that can alter culture drastically. Substantivists attribute two properties to technology. First, they conceive of technological development as something autonomous. Technology gives rise to ever new possibilities, and these new possibilities are always realized eventually. Nothing can stop technological development; technology follows its own dynamic. Second, substantivists ascribe to technology the ability to change culture. Technological development, so to speak, takes culture along with it.

But these two alternatives become problematic within the postphenomenological perspective, for they assume that technology can be spoken about independently of the human beings who engage with it and the culture in which it functions. As Ihde indicates with the concept of "multistability," which was discussed in Chapter 3, technologies are never "in themselves" but are always related to the human beings who engage with them. Technologies receive "stability" only in their use; they are then interpreted as "artifacts-in-order-to . . ." Such multistability exists not only within a single use context, but always has a cultural index as well. One of Ihde's examples concerns the "cultural embeddedness" of sardine cans that developed in New Guinea: in the 1930s these artifacts were left behind by Australian gold prospectors and snatched up by the inhabitants for use as the centerpieces of their headgear (Ihde 1990, 125). In this cultural context, these objects took on another identity.

The multistability of artifacts implies not only that artifacts can have different meanings in different contexts, but also that specific goals can be technologically realized in different ways by a range of artifacts. Ihde's favorite example of this concerns the difference between Western navigational techniques and the traditional navigational techniques of the South Sea islanders. While Western navigation is strongly instrumentally mediated and mathematical in nature—one navigates with charts and compasses—the South Sea islanders navigate by carefully observing

4. For this distinction, see Borgmann (1984, 7–12).

stationary clouds (which hang over islands), birds, and wave patterns (146–49). The South Sea islanders had an extremely complex navigational system and could navigate at least as well as the first Westerners who encountered them at the time of the first voyages of discovery. The South Sea system always "looks" *laterally* from the position of the navigator, which calls for a completely different type of navigation—and is oriented by a completely different set of phenomena—than the Western system, which looks *from overhead,* down on the water and land. Though their navigational system was not technological in nature, instruments could in principle play a role in them, in the form of instruments that could peer through mist, make wave patterns more perceptible, and so forth. The cultural "way of seeing" of the South Sea islanders thus could give rise to an entirely different technology than the Western one involving charts and compasses.

Human ends, therefore, can be realized in many different ways, depending on the cultural context in which they play a role. Different cultural contexts, different "ways of seeing," thus can lead to the development of different technologies. But such multistability does not imply that technologies are only projection screens for cultural frameworks of interpretation. For according to Ihde, technological intentionalities (in the sense of "intentions of technologies") appear at the cultural level, over and above those at the perceptual level discussed in the previous section. In the last part of *Technology and the Lifeworld,* "Lifeworld Shapes," Ihde sketches out a number of character traits of the technological lifeworld to illustrate this point.

The first and most important "cultural intentionality" that Ihde mentions is that technology has transformed our culture into what he calls a *pluriculture.* Here he cites explicitly communication, information, and imaging technologies—technologies that mediate our experience (164–67; see also 1993b, 62). These technologies have made possible an exchange between cultures to such a large extent that they have come to play a role in the everyday life of almost everyone. Thanks to the media, we are confronted with many other cultures than our own. This confrontation does more than allow us to see what goes on in another culture from a distance; it effects an exchange of cultures on a daily basis.

Ihde speaks pointedly not of *multiculturalism* but of *pluriculturalism.* By multiculturalism he understands the coexistence of several cultures that in principle could exist apart from each other. The term "pluriculturalism" denotes, by contrast, that several cultures simultaneously play mutually interwoven roles in our lifeworld. "We all 'pick and choose' amongst the pluricultural fragments, and 'edit' or construct our bricolage life culture" (Ihde 1993a, 64). Pluriculturalism goes further than the fact that we may eat Chinese, Italian, and French foods, and that we may decorate our living rooms with African carvings and Indonesian batiks. The pluriculturalism of the contemporary lifeworld entails that it is not enough to have a single cultural interpretive framework, a single "macroperception." We have to be able to "see" in several ways at the same time; we have to have a "compound eye," in Ihde's words, similar to the way the director of a television program has several television screens playing simultaneously in the studio directing room. The world has become a mosaic and cannot be engaged from a single cultural interpretive framework anymore (Ihde 1993a, 114–15).

A second important transformation in our culture that technological development has wrought is that technologies create a "decisional burden" because of the many new choices they make possible. It is less and less obvious that events or occurrences unfolding now will forever remain what they are because ever more things that hitherto seemed inescapable now are falling under human control, or at least influence, through technological developments. Having children, for instance, is no longer something that simply befalls us but has become a conscious decision. For those who are eager to have children but are unable to conceive, there are a steadily increasing number of technologically assisted options available. Prenatal diagnostics subsequently raises the question if and when it is desirable and morally acceptable to terminate pregnancies; this is but one of the many questions raised by the technologies in question. All of these technological developments create ever more instances, as well as kinds, of choice. And we no longer have the choice to shirk them: "The one choice I do not have is the choice not to make a choice," Ihde says, with a nod to Sartre (Ihde 1990, 181).

Technologies thus possess multistability and intentionality not only at the level of the individual user but also at the cultural level. These concepts form a suitable replacement for the classical opposition between instrumentalism and determinism. Technology is never purely determinative, for in principle other cultural relations with a given artifact are always possible. But neither is it purely instrumental, for when an artifact receives a particular definition within a cultural context—and thus becomes stable rather than multistable—it still contributes to shaping that context. Precisely because technologies are always interwoven with culture, they are always in a position to transform that culture—not "in themselves," but from the position that the cultural definition has given them. When a cultural relation with an artifact is initiated, there arises a "cultural intentionality" within that relation, a cultural space mediated by technology, thanks to which technology is able to give indirect form to the interpretations and experiences of human beings, as well as directly mediating sensory perception.

Science as Technological Hermeneutics

The interpretive framework that has been shaped most profoundly by technological mediation is probably the scientific one. Science plays an increasingly strong role in shaping the ways in which human beings interpret their world. Scientific concepts and theories crop up in specialized journals and professional discourse; they also help shape the way in which, for instance, human beings daily evaluate not only nature but also their own physical and mental well-being. Ihde has devoted much thought to the role of technologies in the generation of scientific theories by connecting his philosophy of technology with the philosophy of science. He does this especially in *Instrumental Realism* and in *Expanding Hermeneutics,* books in which he brings to light an important hermeneutical role of artifacts. For perception—both microperception and macroperception—plays a central role in the connections Ihde draws between his philosophy of technology and the philosophy of science, and in the consequences for the philosophy of science that he sees in his analyses of the role of technology in human perception.

The most obvious connection between Ihde's approach to technology and the philosophy of science arises from his understanding of macroperception. Since the work of Thomas Kuhn, the philosophy of science has developed in such a way as to take ever more seriously the context-dependence of scientific knowledge. In place of seeking possible ways to ground scientific knowledge in reality, to find characteristics of a language adequate to speak about reality, or to discover the conditions of possibility for scientific knowledge, contemporary philosophy of science has sought to understand how the significance of scientific statements arises from the contexts in which they were formulated. Following Kuhn, philosophers of science have seen its development as moving not ever closer toward a final solution to a puzzle, but rather from one framework of interpretation to another. Science always takes place inside what Kuhn called a *paradigm;* it is the work of a community of scientists who share an interpretation of reality as well as a definition of the problems deemed to be important. Michel Foucault did something similar, according to Ihde, through his concept of the *epistèmè.* Foucault's concept is less sociologically laden than Kuhn's conception of paradigms, and describes the "way of knowing" of a specific period specified by the language spoken by its scientists—the "discourse" with which they engage each other (Ihde 1991, 33). In these paradigmatic and epistemic concepts Ihde sees parallels with his own concept of macroperception. Science has to do with the "ways of seeing" of scientists.

This association with "ways of seeing," however, at the same time brings to light a lacuna in contemporary, contextualist ways of thinking about science. Science, to be sure, is to be understood as a "way of seeing," but the discussion thus far has localized this to the macroperceptual level only. Science, however, also has everything to do with "seeing" on a microperceptual level—with concrete sensory perceptions. According to Ihde, science must be related not only to the contexts of interpretation in which it takes place, but at the same time to the sensory perceptions of scientists. And one principal characteristic of contemporary science is that these perceptions are *mediated by technologies.* Right here, according to Ihde, lies the most interesting connection between his approach to the philosophy of technology and the philosophy of science. Philosophers of science have readily acknowledged that scientists "see" reality in a certain way, but have paid insufficient attention to the fact that these ways of seeing are also based on concrete, and technologically mediated, perceptual seeing.

Ihde's broad thought is therefore that the philosophy of science must complement the study of the macroperceptual aspects of science with an analysis of its microperceptual aspects—and the role technologies play in these. In this way, one might say, Ihde gives a new twist to Heidegger's conviction that technology has primacy over science. Technology has primacy, not because the technological mode of thinking is presupposed in scientific thinking, but because contemporary science is helpless without technologically mediated instrumental perceptions (62–63). The mediation of scientific perceptions by technological instruments is no mere accident, but has become an essential part of scientific knowledge. This, in turn, calls for an analysis of the contributions that such instruments make to the knowledge formulated with their aid. After all, from Ihde's analysis of technological mediation, scientific instruments cannot be seen as neutral passageways to "the world itself."

Instruments mediate the perceptions of scientists and transform them in the process. Radio telescopes, for instance, make things "perceivable" that are not accessible to the naked eye. Computer tomographs and ultrasound scanners produce images of the human body and its structures that would otherwise be unobservable. These mediated perceptions therefore reveal entities that we would never have known about but for mediating technologies. Technological instruments, Ihde claims, play an essential role in the generation of scientific knowledge, and studying this role is crucial to understanding contemporary science.

Note that Ihde tries to understand science in terms of what scientists *do,* not just in terms of the structure, conditions of possibility, and foundations of the knowledge they produce. Attending to scientific practice does not lead him, however, to a sociological or anthropological perspective, as it does many scholars in the field of science studies. For Ihde, a turn to scientific practice does not mean a turn to the analysis of the social interaction of scientists, but rather to the embodiment of science in observations and in the instruments with which these observations take place.

A philosopher of science who wants to do justice to scientific practice cannot cling to a naive realism that postulates a one-to-one relation between what scientific knowledge makes visible and what is "really" there. Ihde therefore labels his position *instrumental realism.* The reality studied by scientists is constituted by the technological instruments they use. What interests Ihde most deeply is this connection between instrumental mediation and the content of scientific knowledge, as is evident from the last part of *Expanding Hermeneutics,* whose program is "to show how science can do a 'hermeneutics of things' by turning them into scientific objects" (Ihde 1998, 139). In other words, he asks how we are to understand the scientific way of interpreting reality "in action": how is reality "prepared" by technologically mediated interpretations so that science can be done with it?

For this, the classical meaning of hermeneutics needs to be expanded. Traditionally, hermeneutics was understood to involve the interpretation of texts as well as reflection on the process of interpretation and its conditions. Ihde, however, develops a more material conception of hermeneutics. For him, it is possible to interpret things other than texts hermeneutically, and he even discerns nonlinguistic forms of interpretation, such as those offered by scientific instruments. Scientific instruments constitute what scientists observe; they "interpret" reality before humans can observe it.

Ihde thus adds a material-hermeneutic perspective to the post-Kuhnian thought that science is to be understood as a "manner of seeing." For Ihde, scientific observation is "*through, with, and by means of instruments*" (159). Instruments prepare reality for observation and make scientific objects out of it by making it present in very specific ways. "[T]he instrument is already *a hermeneutic device,*" Ihde concludes (149). Following a path similar to Latour's, wherein scientific objects are prepared in the laboratory in and through the production of "inscriptions" that make things scientifically analyzable, Ihde argues that laboratory instruments make things "readable."

This process of "making things readable" consists of transforming something that is imperceptible or practically so into something perceptible. This can happen via simple magnification, as in the case of microscopes and telescopes, but also by way of more radical means of mediation. For instance, in order to make microorganisms microscopically visible, it is necessary to stain them with aniline dye. In this way they lose their isomorphism with "naked-eye perception"—if unmediated perception of such small organisms is possible at all. Technologies such as X-rays, ultrasound, and MRI scans go a step further: they provide a picture of the human body (or parts thereof) based on the passage of invisible rays, the bouncing of sound waves, or nuclear resonance, phenomena that are not perceivable without technologies and that thus need to be "translated" into the visible.

A much more common way in which instruments make things readable is through what Ihde calls "text-like visualization." Here, instruments create a hermeneutical relation between human beings and reality by providing a representation of reality to be "read": graphs, tables, maps, and so forth. This sort of visualization can suggest an analogy with direct perception, such as the mercury level in a thermometer that is high at high temperatures and low at low temperatures. But such an analogy need not take place. For instance, the way a spectrogram provides an image of a star has no analogy to the star itself.

In principle, two interpretations are possible of the hermeneutical role of scientific instruments. Borrowing the language of constructivism, Ihde characterizes one as a "weak" and the other as a "strong" program. In a weak program, instruments are conceived as forming an interface between science and the reality it studies, codetermining how that reality is to be interpreted. A strong program, which Ihde ultimately advocates, goes a step further and views instruments as actually constituting the objects studied by the sciences and therefore as codetermining the content of scientific knowledge. What an X-ray image or a sonogram makes visible can in principle become visible once an operation is underway. But there also is a class of phenomena that would be unobservable without the help of technological instruments. These phenomena, therefore, are constituted by these instruments.

The investigative domain of astronomy, for instance, was enormously expanded by the development of radio telescopes and techniques to make visible invisible forms of light, such as infrared and ultraviolet radiation. What was not directly observable became constituted as an observable object by translation technologies, allowing new phenomena to be revealed to scientific research. Such phenomena have no equivalent in the perceptible world, and can only become present to us via technologies. In the microscopic world, too, previously invisible realities are made visible by technologies such as the electron microscope. Increasingly things that previously had been imperceptible have become the object of scientific investigation, thanks to their mutual constitution by technologies.

With this strong program, Ihde seeks to make connections, however implicitly, with constructivist approaches to science that have a like vision of the relation between science and reality. Ihde's principal metaphor in describing his strong program is that scientists use their instruments to "give a voice" to things (Ihde 1998, 172), meaning that the way in which the voice is given codetermines what can be heard of it. The technology used constitutes the object that is interrogated without completely determining it.

Neglecting this technological constitution would lead to a new variant of the naive realism that constructivists warn about: the assumption of a correspondence between a scientific theory or observation on the one hand, and "reality-in-itself" on the other (Ihde 1998, 178). This assumption fails to recognize the active, mediating role played by instruments in the coming to be of scientific knowledge. This dimension—that the content of scientific knowledge relates to the context of practices and interpretations in which it arises—deserves more attention than it thus far has received in science studies. However, in so doing, science studies encounters the danger of running into the opposite pitfall of naive realism, namely, naive idealism, or the view that the ultimate font of knowledge is not the world but our ideas about it. Scientific knowledge is a product not only of interpretations, but also of the material conditions on which these are formulated—the instruments with which scientific observations are carried out.

CONCLUSION

Ihde's work allows the possibility of adding a hermeneutic dimension to the postphenomenological approach outlined above, and thus offers an entirely different perspective on technology than that of the traditional hermeneutical perspective developed by Heidegger. For however much both Ihde and Heidegger pose the same question on the matter—"What does technology mean for our relation with the world?"—they arrive at completely different conclusions. Whereas Heidegger sees technology mainly as a controlling way of world disclosure, Ihde articulates a much more nuanced picture of the hermeneutical role of technologies.

The difference between Heidegger and Ihde stems from a difference in the ways in which each conceptualizes technology. Heidegger's approach consisted of investigating the nature of the interpretive relation to reality that makes technology possible. He understood technology to be a way of disclosing meaning, and by extrapolation took this to be *the* way of interpretation toward which Western culture as a whole was headed. Ihde's approach to technology, however, does not begin with this world-interpretation, but rather with our dealings with concrete technological artifacts, and the praxes and interpretations that are made possible by them. When the question of meaning is posed from this perspective, an entirely different picture of technology emerges.

Ihde brings to light the many possible ways in which human beings can relate to technological artifacts: technologies can mediate perception, they can be perceived themselves, and they can be present in the background of human perception while helping to shape it. Mediating relations in particular offer possibilities for taking the postphenomenological perspective in a hermeneutical direction. On the microperceptual level, technology—as a mediator of human perception—does not at all imply a reduction of the ways in which reality manifests itself. Many forms of technological mediation are possible that transform our access to the world in myriad ways, some of which open up to us new ways of access unavailable to "naked-eye perception," and some of which narrow this access.

On the macroperceptual level, technology is no longer seen as fostering an overarching, uniform framework of interpretation in which the world is coerced to appear in the form of Heideggerian standing-reserve. Technological culture does not develop in the direction of one-dimensionality, calculativity, and uniformity, but rather in the direction of plurality (Ihde 1990, 159). Technology does not create one single way of disclosing reality—the "technological way of revealing"—rather, it fosters the proliferation of different ways of seeing within our culture. A glance at technological artifacts themselves is sufficient to show the vapidity of the idea that technological thinking holds our entire culture in its clutches. The cultural role of technologies is far more subtle and ambivalent than that.

Moreover, technology—thanks to its role in scientific perception—plays a constitutive role in the production of scientific knowledge, which is becoming progressively more important in the framework through which human beings interpret the world. The role of scientific instruments in the production of scientific knowledge runs far deeper than that of "depicting reality": instruments mutually constitute the reality investigated. They make visible aspects of reality that otherwise would be invisible, aspects that have to be actively represented and thus "interpreted" by the instrument. In the course of elucidating this process of technoconstitution, Ihde goes much further than Heidegger did in pointing out that science must be seen as applied technology rather than the other way around.

Ihde's philosophy does more justice to technology than the tradition from which it stems. Phenomenology after Husserl has always spoken of its main task as understanding human experience—the world in its everydayness—rather than scientific abstractions of it. With respect to technology, it did not succeed in this. The role of technology in human everyday life involves far more than calculative thinking. Technologies actively shape the relation between human beings and their world, and provide many varied and enriched ways in which that world can be encountered.

5

The Acts of Artifacts

INTRODUCTION

What role do things play in human life and action? How do they contribute to shaping our existence? While in the previous chapter I discussed the hermeneutical dimensions of technological mediation, in this one and the next I shall investigate its existential dimensions. I shall do that in two stages, each elaborating a key aspect. My eventual goal is to analyze the role that artifacts play in the realization of human existence. But in order to do so it is first necessary to elucidate the technical mediations of the actions that give shape to human existence. These two aspects are, of course, closely interrelated: the actions of human beings shape the ways in which they realize their existence, while the form of that existence, in turn, shapes human actions.

The relation between "action" and "existence" in the existential perspective parallels the relation between "perception" and "experience" in the hermeneutical perspective. While the latter sees human experience as arising from the mutual interaction between sensuous perception (microperception) and the context of interpretation that gives meaning to it (macroperception), the former sees human existence as taking shape from the mutual interaction of human actions and the context of existence in which specific ways arise for human beings to engage the world. In order to work out both aspects of existential mediation, I take as a point of departure two starkly different positions. In this chapter I shall investigate mediations of action using the work of Bruno Latour, while in the following one I will investigate mediations of existence, in which the work of the German-American philosopher of technology Albert Borgmann plays an important role. Just as I did in the previous chapter with the work of Don Ihde, I shall critically evaluate and extend the work of these thinkers in order to make them fruitful within the context of my concern to analyze forms of technical mediation.

The decision to engage Latour's work from a phenomenologically oriented perspective might seem surprising at first. The character and ambitions of Latour's perspective and that of phenomenology strongly differ (Latour 1999a, 9).[1] Still, both approaches share the desire to overcome the subject-object dichotomy that, since the Enlightenment, has played such an overarching role in Western philosophy. That common point of departure will make it possible to make use of Latour's actor-network theory within the context of this study.

From Latour's own perspective, translating his concepts into a postphenomenological vocabulary should not be problematic. After all, from this perspective, striving for an "authentic" way to deal with actor-network theory would be a parlous task. "Translations"—such as my reworking of the substance of his work—occupy a central place in this theory. Latour interests himself precisely in the various ways in which entities are continually translated when brought into relation with other entities. In my translation, however, I will take special care to try to bring the two "languages" in close connection with one another, with the aim of producing a fertile hybrid of phenomenological and actor-network theory notions, which can be used to analyze how artifacts coshape human actions.

LATOUR'S AMODERN ONTOLOGY

In order to connect Latour's analysis of the mediating role of artifacts with the postphenomenological perspective, it is important first of all to sketch out the main lines of his actor-network theory as well as the context in which he develops his analysis. As noted in Chapter 3, one main characteristic of Latour's work is the

1. Later on in this chapter I shall have more to say about the relation between Latour's approach and phenomenology.

ambition to overcome the radical dichotomy, so deeply ingrained in modern philosophy, between subject and object. Latour usually describes this dichotomy as a separation between or purification of "humans" and "nonhumans," and he objects to the fact that these have been treated in radically different ways since the Enlightenment. For Latour, reality cannot be adequately understood if humans and nonhumans are treated "asymmetrically." The two cannot be had separately, but are always bound up with each other in a network of relations. Only by virtue of this network are they what they are, and can they do what they do. A point of departure that cleaves to an a priori dichotomy between the human and nonhuman overlooks this mutual involvement and is therefore inadequate.

Actors and Networks

The two most basic concepts that Latour uses to think symmetrically about humans and nonhumans are "actor" and "network." Latour's universe consists of actors that stand in relation to each other and interact via networks. His approach is thus frequently described as "actor-network theory." The concept "actor," however, suggests that the entity that acts, the possessor of "agency," is a human—and if we are to think symmetrically, according to Latour, agency cannot be restricted to human beings. For this reason Latour prefers to speak of "actants" rather than "actors." And, noting that the ensemble of human actors is usually referred to as "society," Latour prefers to refer to associations of human and nonhuman actors as the "collective."

Actants must not be conceived as free-standing entities that then enter into relations with each other. Only in these relations do they become actants; they "emerge" within the networks that exist between them. Were this not so, they would have a pre-established essence, which Latour rejects. The pile of matter that we call an "automobile" can only exist as such in a context that includes also gasoline, gas stations, pumps, refineries, highways, auto mechanics, automobile manufacturing plants, and so forth. What exists "in itself" is only metal and synthetic material.

What counts for Latour are not essences but rather existences (Latour 1997, 24) and, as he puts it with a nod to Sartre, existence precedes essence. In contrast to existentialism, however, Latour sees existence as pertaining also to entities of a nonhuman nature. Existences are the nodes between which networks of relations are strung—still uncongealed and undefined phenomena that only receive form in relations, which temporarily stabilize them and allow them to be taken for granted. Essences are constructed in the networks that exist between existences.

They are the congealed outcomes of the connections and activities in a network that usually remain black-boxed—hidden away and forgotten about.

The reason that Latour insists on treating humans and nonhumans symmetrically in his descriptions is that the roles they play in networks are equivalent. His approach is inspired by semiotics, and can be seen as its material translation. Semiotics deals with the analysis of systems of signs, and views them as meaningful only because first and foremost they refer to each other rather than to a reality "outside" or "beyond" themselves. According to the semiotic approach, the word "book," for example, is not meaningful on the basis of its reference to an object to which it corresponds; what it means takes shape from the connections which this word has with other words like "cover," "page," "text," "letter," and so forth. In a similar way, according to Latour, entities (both human and nonhuman) have connections with each other, which establish not so much meaning but rather a way of being present for human beings.[2] A book presents itself to human beings by virtue of the relations it has with a bookstore, publisher, printer, the tastes of book buyers, and so forth, rather than having some predetermined "essence." According to Latour, what a thing is and what a human being is both arise from their relations with other things and human beings rather than from an "essence" that hides behind them.

This approach might appear to culminate in full-blown relativism. Little of reality seems left over when entities are explained as the contingent outcomes of networks with nothing remaining "in themselves." But this impression is not completely correct. Latour does not deny that a reality exists of which knowledge is possible. In the first chapter of *Pandora's Hope,* for instance, he describes an episode in which a scientist buttonholed him to ask him an urgent question: "Do you believe in reality?" Whereupon Latour laughed and answered, "But of course! What a question! Is reality something we have to believe in?" (Latour 1999a, 1). Latour does not give up on reality; he only refuses to see it as a world "outside of us." Reality is only present by virtue of the relations human beings have with it. What counts as reality for humans is related to the network that connects them with it. In laboratories, for instance, problems are approached in specific ways, terms are defined, disputes pursued, and so forth, all of this constituting a specific network of relations binding researchers and what they are researching together, in virtue of which the latter is present to the former.

Latour is thus expressly not a *social* constructivist who reduces reality to the social interactions of human beings. He is merely a constructivist: he thinks in

2. See Mol and Mesman (1996, 429).

terms of constructions in which both humans and nonhumans play a role. Latour says emphatically that the phenomena observed by human beings are "not our own creations; they are not made out of thin air, not of social relations, not of human categories" (Latour 1993, 25). If knowledge were seen as a social construction, this way of thinking would merely be the mirror image of the naive realism that social constructivism is so concerned to refute. In that case, not nature but the human beings who reflect about nature would become absolutized as the ultimate fountain of knowledge. Scientific facts, according to Latour, cannot be reduced to either humans or nonhumans: they consist of networks between both.

Moreover, Latour says that he has come to prefer to shun the name "actor-network theory" because it has become misleading.[3] This holds true for the concepts represented by the terms "actor," "network," and "theory," as well as the dash between "actor" and "network." One reason for his worry is that the concept "network" has become an everyday word with a completely different meaning than the one he intends. A more important reason, however, is that the concepts "actor" and "network," when separated by a dash, are too reminiscent of the classical sociological "agency-structure" dichotomy, which distinguishes subjects who act from the structure in which that action plays itself out. An "actor," according to Latour, is not a traditional intentional subject and still less is a "network" an ensemble of institutions that provide a framework in which subjects manifest themselves. Actors can be as much human as nonhuman, and networks are not structures but relations in which translations take place of entities that assume relations with each other.

The concept "theory," moreover, suggests precisely the opposite of what actor-network theory seeks to do: to trace entities as they move through networks of relations. Latour says that actor-network theory is "simply another way of being faithful to the insights of ethnomethodology," which inspired his work: "actors know what they do and we have to learn from them not only what they do, but how and why they do it" (Latour 1999b, 19). The pretension to formulate a theory would falsely imply, for Latour, that researchers have reality in hand, so to speak, and would be able to explain to the humans and nonhumans they are researching which laws they fall under without knowing it. This runs directly counter to Latour's ambition to do justice as much as possible to the richness of the ever-changing domain of practices and situations. "Far from being a theory of the social . . . it [actor-network theory] always was, and this from its very inception, a very crude method to learn from the actors without imposing on them an a

3. See Latour (1999b).

priori definition of their world-building capacities. . . . [I]t is a theory that says that by following circulation we can get more than by defining entities, essence or provinces" (20).

Amodernism

As I have already noted, in his analysis Latour seeks to escape from the subject-object dichotomy that has ruled modern philosophy. In his approach to science, this becomes clear in his insistence that neither human beings ("the social," "subjects") nor nonhumans ("nature," "objects") can be seen as supplying the ultimate font and safeguard of knowledge. His position, however, does not boil down to some middle position between realism and social constructivism. If he would position his work this way, he would still take the subject-object dichotomy as a starting point. What, then, does Latour's position amount to? An answer to this question is important in order to localize his analysis of technical mediation.

In Latour's eye, the distinction between nature and society, or subject and object, which has seemed so self-evident since the Enlightenment, needs to be seen as a product of modernity that has far exceeded its expiration date. No other society makes this distinction in such a radical manner, and in ours it is more and more painfully obvious how poorly it allows us to comprehend what is happening in the world. The project of modernity, according to Latour, consists of the attempt to purify objects and subjects—we set objects on one side, subjects on the other, and draw a line between them. What is on the one side of the line is then material for scientists to investigate, with what is on the other side for the social scientists. In so-called premodern societies this distinction was not so hard and fast. In such societies, men were still afraid, as Latour says with a reference to Asterix, that the skies would fall on their heads—a fear that, according to us, who have gone through the Enlightenment, is the product of subjective factors, and is not connected to the objective world outside.

This purification and separation of subjects and objects, according to Latour, is coming to be less and less believable. Ever more entities arise that cannot be comfortably placed in this dichotomy. Latour calls these entities "hybrids." The irony is that these hybrids thrive thanks to the modern purification: precisely because they don't fit within the subject-object schema, we cannot recognize them and therefore they can proliferate at an astounding rate without anyone trying to stop or change them. But now, as their numbers become ever greater, it becomes more and more difficult to deny their existence. We are flooded with entities that straddle the boundary between humans and nonhumans: "frozen embryos, expert systems, digital machines, sensor-equipped robots, hybrid corn, data banks, psychotropic drugs, whales outfitted with radar sounding devices, gene synthesizers, audience analyzers, and so on" (Latour 1993, 49–50). These entities are neither subjects nor objects. "Where are we to put these hybrids? Are they human? Human because they are our work. Are they natural? Natural because they are not our doing" (50). They are subjects and objects, but not pure subjects and still less pure objects. They are human and nonhuman. The ozone hole, for instance, is not a purely natural phenomenon, for it is caused in part by human beings, is made visible only by human beings, and is a subject for concern only because it threatens them. However, neither is it purely human, for it is indeed "something" that can be made perceptible and influence human life. What is going on in the ozone layer, in short, cannot be understood in purely human or nonhuman terms, but requires taking into account the networks of relations between both.

Latour, however, refuses to counter the subject-object dichotomy of modernism with postmodernism. Postmodernism describes itself as something that comes "after" modernism. While recognizing that there is something wrong with the modernist project, it implicitly carries that project forward. The repudiation of the distinction between subject and object leads, in Latour's eyes, to the denial of all connections between them. Science and technology, according to the postmodernists, do not deal with anything human, while reality consists of our own stories and interpretations. Postmodernism thus carries out an extreme purification of humans and nonhumans. Latour does precisely the opposite. His analysis of the hybrid character of the entities in our culture leads directly to the conclusion that, in fact, we have never been modern, however much we may think we have been. The purification process that characterized modernity should be understood as one of the possible ways to deal with hybrids. But it could not prevent the continuous emergence of impure relations between humans and nonhumans.

Modernity acted as if humans and nonhumans were separable, but in so doing it unknowingly affirmed and stimulated their interminglings. Never before have the chains of relations between humans and nonhumans been as long as they are now; never before have humans and nonhumans been intertwined on so great a scale. But this intertwining is invisible to modern philosophy, which can only think in terms of pure humans and pure nonhumans. From a nonmodern or amodern perspective such as that of Latour, however, the process of purification can become visible. By abstaining from the modern craving for purification, it becomes possible to understand not only the existence of hybrids but also the way in which they were made invisible.

A continuity can be seen, in the amodern perspective, between our society and so-called premodern societies. An ethnographer doing field work returns with "a single narrative that weaves together the way people regard the heavens and their ancestors, the way they build houses and the way they grow yams or manioc or rice, the way they construct their government and their cosmology" (7). Why aren't these elements just as interwoven in our own culture? Referring to the gap in the ozone layer, and again to Asterix, Latour says, "We too are afraid that the sky is falling on our heads. We too associate the tiny gesture of releasing an aerosol spray with taboos pertaining to the heavens" (7). Humans and nonhumans are just as bound up together in our culture as they are in others; therefore, Latour concludes, we need to study our technological culture similarly to the ways that anthropologists study other cultures. This means studying how the networks of relations between humans and nonhumans develop and unravel. In order to understand our culture, we must trace out both the process of purification and that of hybridization; we must understand how hybrids arise and why they are not seen as hybrids. In order to understand phenomena, they should be approached as black boxes that, when opened, will appear to contain myriad relations and activity.

TECHNICAL MEDIATION

From his amodern perspective Latour seeks to venture upon the terrain of the philosophy of technology. The philosophy of technology, according to his approach, needs to avoid the same pitfalls that threaten the philosophy of science: the absolutization of the dichotomy between humans and nonhumans. In the philosophy of science, this absolutization takes two directions: realism, which seeks to understand scientific knowledge as originating in the nonhuman alone; and social constructivism, which reduces knowledge to the humans who formulate it. Latour illustrates what these pitfalls look like with respect to technology as follows: if someone shoots another with a gun, who shoots—the person or the gun? "Weapons kill people," say proponents of gun control, while opponents say, "people kill people." The first, materialistic position is based on the view that the gun acts by virtue of the operation of its material components, irreducible to the social properties of the gunman, while the second, sociological position is based on the view that the gunman is the one who acts and that the gun plays a neutral role and does nothing in itself (Latour 1999a, 176–77). Who is right?

The absurdity of this question underscores the necessity of thinking about technology while recognizing the intermingling of humans and nonhumans. No one would claim that the gun does not contribute to the shooting, and still less that it is wholly responsible. Gun control advocates do not hold that guns themselves do the killing; rather, what they mean is that guns affect those who possess them. Gun control opponents do not truly deny that guns play a role in shootings; they mean that a gun is an efficient way of carrying out an act for which other things could also work (176). Guns and human beings are not separate, but intertwined.

Latour makes this intertwining clear on the conceptual level through the concept of "technical mediation." A gun must not be seen as the sole actor in a shooting, and certainly not as a neutral means to an end; there would be no shooting without a weapon, nor without a gunman. The gun plays here a mediating role. It is not simply an intermediary, a neutral object between the gunman and the object, but a mediator that actively contributes to the way in which the end is realized.[4]

How is such technical mediation to be grasped more precisely? Latour identifies four meanings of mediation, each of which I shall elaborate in turn: "translation," "composition," "reversible black-boxing," and "delegation." Each of these highlights a different aspect of mediation, and together they constitute a contribution to the analysis of technical mediation that I wish to formulate. I shall describe each, then critically examine it and make its connections clear to the postphenomenological perspective I am elaborating.

Translation

The first meaning of mediation for Latour is translation. When a technology mediates, it involves the "translation" of a "program of action." What he means can again be illustrated with the example of the gun. Say a person (actant 1) is

4. Latour develops his analysis explicitly in opposition to Heidegger. He opens with an attack on Heidegger's conception of technology—"That Heidegger's interpretation of technology passes as the deepest of interpretations I find surprising" (Latour 1994, 30; see also Latour 1999a, 176)—and regularly returns to the attack. Latour claims that Heidegger is unable to comprehend the phenomenon of technical mediation: "For Heidegger, a technology is never an instrument, a mere tool. Does that mean that technologies mediate action? No, because we have ourselves become instruments for no other end than instrumentality itself." "Heidegger is mistaken" (Latour 1999a, 176), Latour holds, but in stating this he actually shows that he is not acquainted enough with Heidegger's philosophy of technology. His critique only concerns the later Heidegger. As I made clear in Chapter 2, in *Being and Time* Heidegger does make a thorough analysis of tools and equipment and the role that they play in human behavior—an analysis that can be used to augment Latour's.

angry and wants to take revenge on someone, but is not strong enough to do that person physical harm. His "program of action"—the "intention" to take revenge—is blocked. This person, however, can take on a relation with a gun (actant 2). This gun mediates the program of action of actant 1, on the basis of its own program of action—the "function" of shooting. A new actant arises (actant 1 + actant 2), with a new, translated program: the killing of the person against whom one wants to take revenge. The concept "program of action" should be read symmetrically here: it refers as much to the intentions of human beings as to the functions of artifacts, without invoking a distinction between humans and nonhumans on the level at which the terms are applied.[5] The original program of action is thus "translated" or "transformed" in the technical mediation into a new one. Both the gun and the person change in the mediated situation: the person is different with the gun than without, and the gun is different with the person than without. Neither has an "essence"; they have existence, they exist, and they are transformed in their relation to one another.

Composition

Mediation thus consists of making possible a new program of action that arises out of relations that actants have to each other. This means that mediation always involves several actants that jointly perform an action. Responsibility for that action, therefore, is spread out over the ensemble of parts. Latour identifies this complexity of actorship or "agency" as a second meaning of technical mediation, which he calls "composition." By composition he means that action "is simply not a property of humans but of an association of actants" (182). In the example above, it is not the person who shoots, but person plus gun.

Mediation consists, therefore, not only of the translation of programs of action but also and simultaneously of the linkage of actants. These two concepts already give us a rudimentary set of instruments with which to conduct an analysis of technical mediation. Mediation appears to be a matter of hybrids rather than of pure humans and pure objects. These hybrids arise in the form of complexes of humans and technologies. In order to understand these complexes, Latour replaces the dyad of humans and technology with another: substitution and association, or replacement and linkage. These two dimensions span the space in

5. See Latour (1994, 33–34): "The same is true of goals and functions: the former associated more with humans, the latter with nonhumans, but both can be described as programs of action—a neutral term useful when an attribution of human goals or nonhuman functions has not been made."

which technical mediation plays itself out. Association is the dimension in which the forming of compositions is localized; substitution is the dimension of the translations of possible programs of action. Each mediation can be indicated along these two dimensions (Latour 1992, 250–51).

To illustrate these dimensions at work, Latour cites the bulky key rings that hotel managers often attach to room keys to encourage hotel guests to remember to return them when leaving the hotel. The manager might well also hang a sign politely requesting that the guests return the keys upon departure—but the guests might not see the sign or know the language in which it is written. The bulkiness of the key ring saves the manager endless trouble, and relieves guests of the responsibility of worrying about the matter, for it is simply too inconvenient not to return it.

This situation cannot easily be understood in terms of a radical dichotomy between humans and nonhumans. The key ring is not present here as a thing in itself possessing an essence, and neither are the humans who use it. The key ring has been made deliberately bulky with an eye to its future use by hotel guests, and they in turn are constrained by the key ring—it transforms their desires so that they want to return it on departure. In Latour's own words: "The bizarre idea that society might be made up of human relations is a mirror image of the other no less bizarre idea that techniques might be made up of nonhuman relations" (239). "You could as well imagine a battle with the naked bodies of the soldiers on the one side and a heap of armors and weapons on the other" (Latour 1997, 77).

The role of the bulky key ring can readily be understood in terms of the mediation of programs of action. The hotel manager's program of action is: Make the guests return their keys upon departure. This program of action perhaps conflicts with that of the guests, which has the role of an "antiprogram": they are not necessarily inclined to return their keys. If the manager only wishes the guests to turn in their keys without doing anything specifically to encourage this, the dimension of association (the "and" axis) with respect to his program of action only contains the manager and keys, and with respect to the antiprogram all hotel guests (see figure 2, situation 1).[6] The manager, however, can forge a connection with another entity. When he does so, we are shifted along the dimension of substitution (the "or" axis). The owner can add an oral message such as "Please turn in your room keys upon departure" (2), he can hang up a sign to that effect (3), or he can attach a large and weighty object to the key (4). With each of these associations the manager can broaden his network in order to realize his program

6. This figure is taken from Latour (1997, 58).

of action. At the same time, these associations transform the behavior of the hotel guests, and make the number of actants in the antiprogram shorter—though new actants may join the antiprogram, such as the dog who sees the key ring as a toy and therefore wants to make away with it.

Reversible Black-Boxing

By couching his analysis in terms of translation and composition, Latour is able to avoid the dichotomy between subjects and objects, the social and the technical world. Mediation consists in the blending of humans and nonhumans. This blending, however, usually remains hidden, for it is subject to "reversible black-boxing," Latour's third meaning of mediation. According to Latour, black-boxing is "a process that makes the joint production of actors and artifacts entirely opaque" (Latour 1999a, 183), a process that renders invisible the network of relations that contribute to the entity.

As an illustration, Latour appeals to the example of an overhead projector. During a lecture, he says, this device serves as an intermediary, completely determined by its function and not itself noticeable. But once it breaks, Latour says—wholly but unintentionally in line with Heidegger's analysis of the tool—we are forcibly reminded of its existence. The network of relations in which it participates immediately becomes visible; suddenly all kinds of people and artifacts materialize: repairmen, light bulbs, lenses, screws, and so forth. Until that moment, these entities were invisible parts of the "black box" of the projector (183). Latour invites us to contemplate the time past when these entities existed separately, before the glass, metal, and other raw materials were brought together in this particular overhead transparency.

> Most of these entities now sit in silence, as if they did not exist, invisible, transparent, mute, bringing to the present scene their force and their action from who knows how many millions of years past. They have a peculiar ontological status, but does this mean that they do not act, that they do not mediate action? Can we say that because we have made all of them—and who is this "we," by the way? Not I, certainly—should they be considered slaves or tools or merely evidence of a Gestell? (185)[7]

7. This example reveals, once again, Latour's ignorance of Heidegger's early philosophy of technology. Latour's claim that an artifact only changes over from an invisible mediating role to explicit awareness in breakdown situations corresponds exactly to Heidegger's description of how tools change over from being ready-to-hand to being present-at-hand when they break.

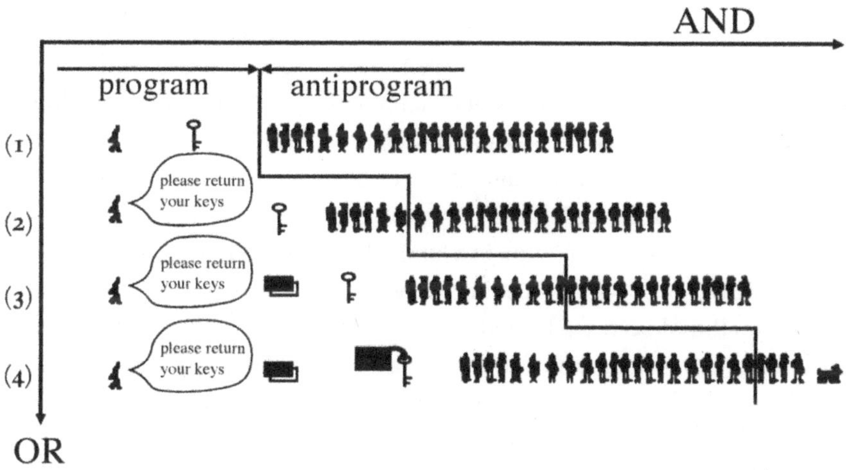

Fig. 2 Mediation, association, substitution

Latour uses the concept of reversible black-boxing—which he also describes as "the folding of space and time"—to make clear that the intermingling of humans and nonhumans is usually invisible to human beings but nevertheless does at times come to light. Moreover, this intermingling occurs not only when connections are made between actants to make possible the realization of a program of action, but also in the separate elements of the chain of connections. The connections a manager makes with a key ring in order to encourage the guests to turn in their keys can be black-boxed, but the key ring, in turn, also consists of black-boxed connections between humans and nonhumans.

Delegation and Scripts

Latour calls his fourth and final meaning of technical mediation "delegation," which he also says is the most important (185). Here he uses the example of a speed bump on a university campus, which translates a driver's goal from "slow down so as not to endanger students" into "slow down in order to protect your car's suspension." Thus drivers adapt their behavior thanks to the mediation of the speed bump. But what is involved here is not only the transformation of a program of action, but also a change of the medium of expression. Drivers now go slowly not because they have read a traffic sign or because they fear a policeman, but because of a lump of matter. Engineers have "inscribed" the program of action they desire (to make drivers slow down on campus) in concrete, as it were. Latour deliberately uses the word "inscribe" rather than "objectify" or "materialize" in order to avoid the impression of "an all-powerful human agent imposing his will on shapeless matter" (186). He is concerned to avoid introducing a new dualism between humans and nonhumans. "We have not abandoned meaningful human relations and abruptly entered a world of brute material relations. . . . [W]e remain in meaning but no longer in discourse; yet we do not reside among mere objects" (187).

Inscribing a program of action in a lump of concrete thus delegates the task of a traffic sign or a policeman—getting people to slow down—to the speed bump. Delegation thus deserves to be understood as a "shifting" or displacement (187). Such shifting has several dimensions: actorial (a bump is not a policeman), spatial (it is located in the middle of the road), and temporal (it is there day and night). Neither the policeman who otherwise would have to be on duty, the traffic sign that otherwise would instruct drivers not to drive fast, nor the engineer who designed the bump are on the scene. Delegation makes possible a curious combination of presence and absence: an absent agent can have an effect on human behavior in the here and now. "Think of technology as congealed labor. Consider the very notion of investment: a regular course of action is suspended, a detour is initiated via several types of actants, and the return is a fresh hybrid that carries past acts into the present and permits its many investors to disappear while also remaining present" (189).

Latour utilizes the example of a door-spring to work out further the process of articulation or inscription (Latour 1992, 226–36). Humans delegate to the door-spring the task of shutting the door after somebody opened it; they inscribe the program of action "close the door if it is open" in the spring. The door-spring invites in turn a particular kind of use. If, for instance, it is made with a strong spring, the person who yanks it open runs a risk of getting slammed in the face by the return swing of the door. The door-spring therefore promotes a particular way of opening a door and passing through. Technologies, as it were, can implicitly supply their own user's manuals. They coshape the use that is made of them; they define actors and relations between actors, and share responsibilities and competencies between humans and things. Latour calls the behavior that a "nonhuman delegate" imposes on humans a "prescription." Prescriptions can be characterized in terms of a series of imperative expressions: "Don't swing the door open too hard," "Walk quickly through," "Don't walk right behind someone else passing through the door," and so forth. Latour indicates such "built-in" prescriptions as the script of a technology (259–60). A script is thus the program of actions or behavior that an artifact invites, expressed in words similar to the series of instructions of a program language.

The designer of an artifact thus works with an inscribed user in mind, to whom he prescribes properties and behavior. This does not mean that users automatically act exactly in the way the designer intended; they have to subscribe to the inscriptions (236). They can simply refuse to use the artifact, or use it selectively and even in novel and unexpected ways—a phenomenon that Ihde referred to through his concept of multistability. But it is often possible to anticipate the behavior of users by taking into account specific groups of users with specific properties. Latour calls such anticipation preinscription (237).

These four meanings of mediation are closely interrelated. In the case of the speed bump example, for instance, this interrelation can be formulated as follows: the president of the university campus where the speed bump has been installed associates himself with a lump of concrete (composition), assigning to it what is necessary to realize his goal (delegation). The resulting speed bump does not need the president or its constructors to fulfill its task (black-boxing) because its physical properties allow it to change a driver's program of action from "drive slowly to be responsible" to "drive slowly to protect my shock absorbers" (translation). Translation, composition, reversible black-boxing, and delegation each illuminate a different aspect of technical mediation.

ACTOR-NETWORK THEORY AND POSTPHENOMENOLOGY

Latour's analysis uncovers a completely different form of technical mediation than Ihde's. While the latter shows how artifacts coshape human experience, Latour illuminates that human behavior can be technologically mediated as well. Thus his approach covers precisely that dimension of the relation between humans and world that in the postphenomenological perspective is complementary to the hermeneutic dimension that was worked out in the previous chapter. Latour's concepts, however, cannot straightaway be used in the context of a phenomenologically oriented approach. We need to translate them into another vocabulary, and to do this in an adequate way requires a closer examination of the mutual relations between each.

A first problem in translating Latour's vocabulary is generated by his express aversion, mentioned at the beginning of this chapter, to the idea that his work could be congealed into a theory. Latour's ambition consists in tracing out the ways in which worlds are built out of collectives of humans and nonhumans, not in the determination of the laws governing these world-building activities. "It is us, the social scientists, who lack knowledge of what they do, and not they who are missing the explanation of why they are unwittingly manipulated by forces exterior to themselves and known to the social scientist's powerful gaze and methods" (Latour 1999b, 19). By taking up his concepts in a postphenomenogical analysis of mediation by artifacts, I place them in a framework that indeed amounts to a systematic theory. And while Latour insists that the vocabulary used "cannot be poor enough" (20), my aim is to construct a vocabulary that is as rich as possible.

The difference between the two approaches and vocabularies, however, should not be overestimated. The ambition of the postphenomenological perspective is in no way to formulate a theory that aims to "explain" empirical reality. My ambition is not to seek out laws that reality obeys, but rather to find concepts with which to make visible and understand as many aspects of reality as possible. Without a suitable conceptual framework, no descriptions—and no actor-network theory—would be possible. Latour says that he is "faithful to the insights of ethnomethodology," by beginning with actants and not with theories about them (19). But ethnomethodology is an approach to cultural anthropology that has been strongly influenced by phenomenology, and Latour cannot possibly be open to accepting the legacy of classical phenomenology with its ambition to "describe" the world itself (in Merleau-Ponty's words; see the beginning of Chapter 4). Descriptions are never impartial; they are mediated at the very least by the vocabulary in which they are couched. And my working out of a postphenomenological perspective aims at a careful building up of such a vocabulary.

A more serious objection to the application of Latourian concepts to my project is Latour's declaration that phenomenology is incompatible with his position. According to him, phenomenology does not overcome the dichotomy between humans and world but rather consolidates and reaffirms it. In *We Have Never Been Modern,* he makes no secret of his objection to phenomenology. In the framework of a discussion titled "Philosophies Stretched Over the Yawning Gap"—by which he, of course, means the gap between humans and nonhumans—he writes:

> Again, one last time, phenomenology was to establish the great split, but this time with less ballast: it jettisoned the two poles of pure consciousness and pure object and spread itself, literally, over the middle, in an attempt to cover the now gaping hole that it sensed it could no longer absorb. Once again the modern paradox is taken further. The notion of intentionality transforms a distinction, a separation, a contradiction, into an insurmountable tension between object and subject. . . . The phenomenologists have the impression that they have gone further

> than Kant and Hegel and Marx, since they no longer attribute any essence either to pure subjects or to pure objects. They really have the impression that they are speaking only of a mediation that does not require any pole to hold fast. Yet like so many anxious modernizers, they no longer trace anything but a line between poles that are thus given the greatest importance. Pure objectivity and pure consciousness are missing, but they are nevertheless—indeed, all the more—in place.[8] The "consciousness of something" becomes nothing more than a slender footbridge spanning a gradually widening abyss. Phenomenologists had to cave in—and they did. (Latour 1993, 57–58)

Latour is partly right, but he doesn't see far enough around the bend. Phenomenologists share his goal to overcome the subject-object dichotomy. Both approaches are convinced that the radical dichotomy between subject and object that informs modern philosophy is insupportable, and both are opposed to the absolutization of these poles in the form of realism or naturalism on the one hand, and idealism or social constructivism on the other. In the eyes of each approach, subjects and objects are lined with each other, and each approach thinks primarily in terms of relations and is willing to cast itself as "relativistic" or "relationistic" in order to make clear the importance of these relations.

It must be admitted, though, that in wanting to overcome the dichotomy between subject and object by referring to their mutual engagement, classical phenomenology does not deny the existence of the poles but takes them as the point of departure for its analysis. That problem, however, is overcome in the postphenomenological perspective pursued in this work. This perspective comes much closer to Latour's perspective on the relations between humans and nonhumans. It deals with subjects and objects, not as pregiven entities that assume relations with each other, but as entities that are constituted in their mutual relation. Human beings are what they are by virtue of the way in which they realize their existence in their world, and their world is what it is by virtue of the way in which it can manifest itself in the relations humans have to it. Thus postphenomenology does not draw a line between two poles, but rather lets the poles emerge from the line that constitutes them.

The postphenomenological perspective is therefore no more a reaffirmation of these two poles than the existence of networks and hybrids is an affirmation of

8. The translators have chosen to render this sentence and the preceding one quite freely: "Et pourtant ils ne dessinent plus qu'un trait entre des pôles réduits à presque rien" (Latour 1993, 79).

the separate existence of subjects and objects. Relations between subjects and objects, considered phenomenologically, can only be understood with the help of the two poles "subject" and "object," but in each moment that the relations actually exist the two poles are already intertwined. And that agrees with Latour's vision of the relation between humans and nonhumans as he states it in *Pandora's Hope:* "There is no sense in which humans may be said to exist as humans without entering into commerce with what authorizes and enables them to exist (that is, to act)"; "A forsaken gun is a mere piece of matter, but what would an abandoned gunner be?"; "Objects that exist simply as objects, detached from a collective life, are unknown, buried in the ground"; "Objects and subjects are made simultaneously" (Latour 1999a, 192, 193, 196).[9]

But even though both positions deny the existence of "pure" objects and "pure" subjects, yet the "existence" (in Latour's sense; see Latour 1997, 24) of subjects and objects or of actants must be accepted to make sense of human-world relations or networks. Without this, they would have the same relation to each other that Baron Von Münchhausen had with his scalp: when they would bring themselves into being from out of nothing, this would be as miraculous as saving yourself and your horse from a quagmire by pulling on your own hair. Subject and object constitute each other not from nothing, but by virtue of their existence, understood as a transcendental construction: existences are not empirically perceivable but yet they are a necessary presupposition—inaccessible though they be—in order to account for the mutual constitution of subjectivity and objectivity. Subjects and objects are not building stones between which human-world relations are built up, but rather products of these relations, just as existences can only become actants in concrete networks of actants.

That the postphenomenological perspective does not proceed from pregiven subjects and objects, however, does not eliminate the fact that it always thinks in terms of a distinction between the two, while that distinction is not made in actor-network theory, which sees actants as both human and nonhuman. To properly evaluate the relation between the two positions, the most relevant question, therefore, is not whether but how they overcome the subject-object dichotomy. Actor-network theory is primarily interested in unraveling the networks of relations by virtue of which entities emerge into presence, while a postphenomenological approach, by contrast, seeks to understand the relations

9. Cf. Ihde (1993b, 34): "Were technologies merely objects totally divorced from human praxis, they would be so much 'junk' lying about. Once taken into praxis one can speak not of technologies 'in themselves,' but as the active relational pair, human-technology."

that humans have with those entities—and for which the network of relations and interactions that allows the entities to emerge into presence is not the primary focus of interest.

From a Latourian perspective, phenomenology always concerns itself with a particular network, a specific "chain of associations," namely, that between a human and a nonhuman or a human and another human. In a postphenomenological perspective on the philosophy of technology, a third entity enters this "chain" or "network": an artifact that mediates the human-nonhuman or human-human relation, and thus also the way in which both are constituted.

Latour toys with conceptualizing such chains of associations, notating humans as "M" and nonhumans as "NM."[10] Following this notation, the technologically mediated relations between humans and world (the focus of postphenomenological attention) could be indicated as "M-NM-NM" or "M-NM-M." When the two approaches are compared in this way, the difference between them appears to consist primarily in the length of the chains of relations with which they are concerned. In comparison with actor-network theory, the postphenomenological focus on only two chains of association appears to constitute a much more limited repertoire. While Latour in principle can study endless numbers of chains, postphenomenologists seem to be restricted only to two—and must stand by helplessly if they encounter chains without humans or with more than three elements. But the difference between the two approaches is more subtle than that, for in these short chains the postphenomenological perspective can bring to light things that remain invisible to actor-network theory.

The postphenomenological perspective, for instance, offers a more nuanced look at the connections between the entities in its chains. Latour views these connections simply as "associations," as a kind of cement between actants. Because of that cement, the actants can either act collectively (gun + human = gunman), or collectively make another entity emerge (the computer in my home office is the outcome of a network of software and hardware manufacture, my university, my colleague's car in which he brought it to my house, and so forth). The postphenomenological perspective approaches the connection between the short networks it examines in a more detailed way: in terms of experience and behavior,

10. He writes, "We could view in each chain, of course, the old classification system of the Moderns. M-M-M-M-M would then correspond to 'social relations,' NM-NM-NM-NM would resemble a 'machine,' M-NM the human-machine interface, NM-NM-NM-NM-M the effect of a technology on a human, M-M-M-M-NM the 'social influences on technology,' M-M-M-NM-M-M-M the manufacture of a tool, while NM-NM-NM-MNM-NM-NM would resemble those poor human beings crushed by automations" (Latour 1997, 39).

readiness-to-hand and presence-at-hand. Much more can be said, for instance, about the relation between a human and a computer than that there exists an "association" between them. This association is of a particular kind, in which the computer is present as ready-to-hand, and makes possible new associations with other entities. The latter associations have, on the one hand, a behavioral dimension—my computer and I are jointly writing this text—and, on the other hand, a dimension of experience that does not explicitly surface in the interactional universe of actor-network theory—thanks to my computer I can, for instance, encounter and experience humans via email.

Moreover, the chains of associations studied by the postphenomenological perspective are longer than they appear. For their endpoints and the mediating actants within them can be treated as what Latour would call a black box, containing in principle all the chains that actor-network theory sets about studying. The humans who experience and act, the mediating artifacts, and the humans and things that are experienced or dealt with—all these can be seen as the ends of networks allowing them to be present. For postphenomenology these black boxes can remain closed, however. Not because this approach lacks respect for actor-network theory—quite the contrary—but because it asks a different kind of question. It is interested not so much in the networks of relations on the basis of which the mediating artifacts and the experiencing humans are present, but in the nature of the relations that human beings—thanks to these artifacts—can have to other humans and things.

The difference in the ways the two positions approach the modern dichotomy between subject and object flows from just this difference in the questions they ask. Postphenomenology and actor-network theory want to do away with the gap between subject and object, but their different perspectives ensure that they do so in different ways. Latour claims that the gap does not exist at all and thus that there is nothing to overcome, for within his perspective the difference between subject and object is untenable; they appear to be the same from the perspective of how they appear in networks. Phenomenology and postphenomenology bridge the gap rather than denying it, by bringing to light the mutual engagements that constitute subject and object. Their perspectives are focused on the relation between humans and their world and, contra Latour, do not look "from an externalist perspective" to describe how configurations of humans and nonhumans are continually arising everywhere. And from their perspectives it is indeed meaningful to make a distinction between someone who experiences and something that is experienced, someone who acts and a world in which action takes place—regardless of how interwoven and mutually constituted they are. In human experience, that difference is a vivid reality: each approach is aware of the difference between "self" and "other," between "other humans" and "things." A philosophy that is to have any hope of grasping the everyday reality of human beings needs to acknowledge this difference and to show that making a distinction between them can never imply an actual separation. Contrary to what Latour suggests, therefore, postphenomenology and actor-network theory are not mutually exclusive but complement each other, and Latour's caustic rebuke of phenomenology in *Pandora's Hope* is misplaced:

> Phenomenology deals only with the world-for-a-human-consciousness. It will teach us a lot about how we never distance ourselves from what we see, how we never gaze at a distant spectacle, how we are always immersed in the world's rich and lived texture, but, alas, this knowledge will be of no use in accounting for how things really are, since we will never be able to escape from the narrow focus of human intentionality. Instead of exploring the ways we can shift from standpoint to standpoint, we will always be fixed in the human one. . . . For all its claims to overcoming the distance between subject and object—as if this distinction were something that could be overcome! As if it had not been devised so as not to be overcome!—phenomenology leaves us with the most dramatic split in this whole sad story: a world of science left entirely to itself, entirely cold, absolutely inhuman; and a rich lived world of intentional stances entirely limited to humans, absolutely divorced from what things are in and for themselves. (Latour 1999a, 9)

Latour's critique of the classical phenomenological dichotomy between a "cold, absolutely inhuman" world of science and a "rich lived world of intentional stances" is correct. Discontent with precisely this dichotomy, and with the alienation diagnosis that underlies it, is precisely what drives me to a postphenomenological perspective. Latour correctly says that, from a phenomenological perspective, human beings "will never be able to escape from the narrow focus of human intentionality"—though that focus is broader than Latour thinks, because all experience and action that fill a human life from minute to minute are continually passing through it. In their everyday relation with their world human beings can never abandon their "human standpoint." But it is wrong, from the perspectives of both phenomenology and postphenomenology, to claim that human beings are "absolutely divorced from what things are in and for themselves." Human beings are always with things: this is precisely what intentionality means. It is not at all a problem for a phenomenologist to acknowledge that the objectivity of things and the subjectivity of human beings emerges thanks to networks of relations. This insight even comes as a welcome contribution. The network of relations that allow entities to emerge into presence, however, is something different from the network of relations that humans have with these entities.

In the latter instance one looks "from inside" to the relation between humans and world (that is, from the experience of an action in the world); in the former "from outside." Latour argues not from the standpoint of human beings who are concretely situated in the world, but from the standpoint of an analyst who describes configurations equally from the perspective of humans and nonhumans. This allows him to "shift from standpoint to standpoint," instead of "always being fixed in the human one." What postphenomenology contributes to actor-network theory is the situated perspective, the perspective "from inside out," thanks to which part of the perceived associations and translations can be more closely analyzed in terms of experience and action, existence and meaning, readiness-to-hand and presence-at-hand. Correspondingly, actor-network theory contributes to postphenomenology a way to elucidate the networks of relations that allow entities to be present. And it also provides a vocabulary with which to analyze the mediation of action, which is my main interest in this chapter.

MEDIATION OF ACTION

Now that it is clear that the postphenomenological perspective and Latour's actor-network theory are not as incompatible as Latour himself supposes, the way is open to forging from these two approaches a fruitful way to analyze technological mediation. To do so, I shall seek to translate Latour's vocabulary—which calls for some care, of course, since even the concept of mediation has a different meaning in each vocabulary. In the postphenomenological perspective, artifacts mediate human-world relations, while in actor-network theory they mediate relations between actants in networks. An adequate translation of Latour's vocabulary into a postphenomenological philosophy of technology thus calls for an investigation of what concepts such as composition, translation, blackboxing, and delegation can mean for analyzing the short network consisting of humans-artifact-world, and in particular with respect to the dimension of action within this network.

In translating Latour's vocabulary, I can be brief with respect to the concepts of association and black-boxing. The chains of association with which the postphenomenological approach to technological mediation is principally concerned are of a quite specific kind; namely, those involving a human-artifact-world relation, in which the associations involved in the chains are described in a more differentiated manner than Latour is able to do—in terms of readiness-to-hand and presence-at-hand, experience and action. The availability of this analytically richer alternative makes it unnecessary to include the concept of association in a postphenomenological vocabulary. Black-boxing, too, does not require any special attention in a postphenomenological perspective. Human beings, mediating artifacts, and the entities in the world that are experienced and encountered can all be viewed as black boxes made possible by hidden networks that allow them to be present—but this focus on the origins of entities is not primarily relevant in a postphenomenological perspective.

The concept of delegation, however, requires a deeper elaboration. Latour means by this concept the way that an actant assigns responsibilities and competencies to another actant. But a danger lies hidden in the way he elaborates this thought; namely, an asymmetrical treatment of humans and nonhumans. And from the postphenomenological perspective, this asymmetry fails to do justice to the mediating role of artifacts. The examples of delegation Latour cites generally involve delegations from humans to nonhumans. House owners give to doorsprings the task of ensuring that the door gets shut; hotel managers give to bulky key rings the task of making sure that guests return their keys; and engineers and university officials give to speed bumps the task of getting drivers to slow down. In these examples, Latour seems to reduce what nonhumans do to what humans delegate to them. The symmetry that his analysis possesses on a conceptual level is much less clear on the level of empirical examples. This impression is reinforced when he speaks of the "inscription" of responsibility or morality in things, for that concept—not to mention "delegation" itself—suggests a goal-directedness most often associated with humans.

Moreover, the concepts of "delegation" and "inscription," together with the thought that things contain a "script," invite us to treat the mediating role of artifacts as a property of the artifacts themselves. In the previous chapter, however, it became clear that in the postphenomenological perspective mediation needs to be localized in the relation between humans and world; "technologically mediated intentionality" was to be treated not as a property of the mediating artifact, but rather as a mode of the intentional relation between humans and world. The mediating roles of artifacts depend in part on the kinds of relations in which they are involved: they are always "multistable," to use Ihde's term. When their mediating role is viewed as a property of the artifacts themselves, it invites an asymmetrical treatment that makes it possible to see this role as something "inscribed" or delegated by humans to nonhumans.

This asymmetry is nothing more than a threat, however. Implicitly, Latour does speak of delegations from nonhumans to humans, without explicitly indicating them as such. In his article "Where Are the Missing Masses?" he expresses admiration for the hydraulic door pull: "Especially clever is its way of extracting energy from each unwilling, unwitting passerby" (Latour 1992, 233–34). The pull thus delegates to humans the task of supplying the energy needed to close the door tightly after it is opened. Creating conceptual space for delegations by nonhumans to humans is highly important, for it makes it possible to observe more in artifacts than only what is delegated to them, or inscribed in them, by humans. In many cases, that is, things do much more than what humans intend. To a revolving door is delegated the task of keeping out the draft without making a building inaccessible; that it also makes entering in a wheelchair next to impossible was never an explicit intention.

Such delegation of things to humans is of great interest within the postphenomenological perspective. Delegations in the other direction can remain black-boxed, because for understanding human-artifact-world relations only the mediating roles of artifacts themselves are relevant, rather than their origins.[11] Delegations by nonhumans to humans, on the contrary, form one of the ways in which artifacts can mediate the actions of humans. When delegation is described as the involvement of actants in the realization of programs of action, the delegation by things to humans can be defined as the involvement of humans in the way things perform their programs of action. In this way they give form to the actions of human beings and their involvements with the world, as I shall elaborate further in the next chapter.

The concept of translation, finally, plays the most important role in the context of this study, for it offers the possibility of bringing to light what mediation can mean in the domain of action. The meaning of this concept is shifted somewhat when it functions in a postphenomenological vocabulary. In Latour's perspective, artifacts translate programs of action. When a hotel manager undertakes an asso-

11. When the design process anticipates this mediating role of products and explicitly tries to "build in" forms of mediation in artifacts, this concept again becomes important, as will become clear in Chapter 7.

ciation with a bulky weight, his program of action is translated from "nag about the keys" to "attach a weight to the keys." The hotel guests also undertake an association with the weight, or they could not have a way to enter their rooms. In their case a different translation occurs, from "be responsible and turn in the key when you check out of the hotel" to "get rid of that awkward thing attached to the key as soon as possible."

In the postphenomenological perspective the concept "translation" does not describe the reformulation of programs of actions but rather the mediation of the action itself. The postphenomenological perspective analyzes a specific chain of associations between actants, to put it in Latourian terms: the mediated relation between humans and world, of which action is one dimension. When hotel guests have the room key in their possession, the bulky attachment mediates the way in which they experience the key—namely, as something annoying—and therefore the way they deal with the key—namely, returning it to the reception on their departure. In the latter case, there is a translation of action, a displacement or transposition of the way in which humans are present in and deal with their world. Artifacts coshape the use that is made of them, and thereby the relations that arise between humans and their world. In this mediation emerges a translation of action, which parallels the transformation of perception discussed in the previous chapter.

To develop further the concept of translation in a postphenomenological context, the parallel with the hermeneutical concept of transformation can be examined more closely. In the analysis of hermeneutical mediation it became clear that the transformation of perception possesses a specific structure, which consists of amplification and reduction: particular aspects of reality are strengthened, while other aspects are not accessible or become more difficult to access. A similar structure can be discerned in the translation of action: artifacts invite particular actions while discouraging others or even rendering them impossible. In place of amplification and reduction, then, one might speak of invitation and inhibition. A bulky key ring promotes the return of the key at checkout time and discourages guests from taking it with them. A speed bump inhibits fast driving and invites slow speeds. Via such translations, artifacts mediate the constitution of objectivity and subjectivity just as in the hermeneutic perspective. They make possible particular praxes and in so doing they shape the relations between humans and their world. Because mediated actions make humans encounter the world in a particular way, the mediating artifact helps to determine how both the world ("objectivity") and those who act in it ("subjectivity") are present.

CONCLUSION

The theories of Ihde and Latour help put flesh on the structural skeleton of technological mediation outlined in Chapter 3 in the domains both of praxis and experience. Technological artifacts appear to be more than functional instruments. When functioning, they appear to be present for human beings in a specific way. They hide themselves in the relations between humans and world, and from their "withdrawn position" they actively shape these relations by transforming both experience and action. The way they do so involves amplification and reduction, invitation and inhibition. In this way, they coshape both the way human beings are present in their world and the world is present for human beings.

The hermeneutical perspective directed attention not only to the role that artifacts play in human perception (microperception), but also to the way in which these perceptions acquire meaning—which is coshaped by the macroperceptual frameworks in which human beings find themselves. This distinction can be paralleled in the existential framework by distinguishing between human action, which was the principal theme of this chapter, and the specific forms of engagement or involvement with the world that arise from such action. This engagement with the world, as made possible by actions, shapes human existence, just as, within the hermeneutical perspective, interpretations on the basis of perceptions shape human experience. The question of how technology mediates the involvement and engagement of human beings with their world and the ways in which they realize their existence demands an approach other than Latour's. Accordingly, in the next chapter I shall elaborate such an approach through a critical evaluation of the philosophy of Albert Borgmann.

6

Devices and the Good Life

INTRODUCTION

After having analyzed the structure of the technological mediation of action with the aid of Latour's philosophy of technology, my concern now is to show how artifacts also simultaneously coshape human existence. I shall use the work of the German-American philosopher of technology Albert Borgmann as a point of departure. Borgmann has written extensively about the role of technology in shaping human existence and in the definition of "the good life." He pays particular attention to the way technology shapes the involvements of human beings with their surrounding world. In his approach, these involvements are what is most relevant to study when investigating human existence.

Borgmann was strongly influenced by Heidegger. Indeed, his philosophy of technology can be read as an extension and elaboration of "The Question Concerning Technology" and "The Thing."[1] But even though I counted Heidegger's

1. See, for instance, Borgmann (1992a). He speaks there of "the fruitful field of inquiry that is marked out by the constellation of the framework of technology [the *Gestell*—PPV] and the fourfold of centering things [the *Geviert*—PPV]" (138). At the same time, however, Borgmann indicates that he is less interested in analyzing the history of being and much more interested in everyday human existence and the way in which that is shaped by technology.

philosophy of technology as part of the hermeneutical perspective, it is very well possible to use Borgmann's work to elaborate the existential perspective, as will become clear. However, his work, too, must be treated critically, inasmuch as his Heideggerian inspiration sometimes brings him into proximity with the diagnosis of alienation that infects much classical philosophy of technology. Nevertheless, Borgmann makes a welcome and new contribution to the tradition from which his work springs. In his analysis of technology, he directs his attention not to "Technology" as a monolithic force, but rather to specific technological artifacts. He addresses the question of the role that technology plays in the shaping of human existence, by reducing it neither to a *Gestell* nor to a kind of functional and bureaucratic outfitting of society, as Heidegger and Jaspers did, but by investigating the patterns that technological artifacts give rise to in human life.[2]

THE DEVICE PARADIGM

In *Technology and the Character of Contemporary Life*, Borgmann distinguishes between three different kinds of approaches to technology—substantivism, instrumentalism, and pluralism—and situates his philosophy of technology with respect to each. *Substantivism* views technology as an independent power that unfolds according to its own logic and that holds society and culture firmly in its grasp. An emblematic representative of this approach is Jacques Ellul, who wrote of what he called the *autonomy* of technology. For Ellul this means that human beings have lost control of technology and that it holds them in its clutches. What Borgmann finds problematic in this approach is its inability to explain *why* technology is seen as developing autonomously. According to Ellul, technology follows a logic of efficiency, but at the same time efficiency is said to be the hallmark of technology. This is a circular argument according to Borgmann: technology is put forward here as "its own unexplained explanation" (Borgmann 1984, 9).

Instrumentalism is in many respects the opposite of substantivism, for it sees technology not as independent but as neutral, a mere means for the realization of human ends. Borgmann finds this explanation unconvincing as well, because the role it assigns to technology is too meager. Technology has drastically altered culture and human life—and insofar as it can indeed be understood as a neutral

2. For an extensive introduction to Borgmann's work, see Tijmes (2001). A critical discussion of Borgmann's philosophy can be found in Verbeek (2002, 69–92). This article follows the same line of argument as the present chapter, but includes Borgmann's later work on information technology.

means, instrumentalism glosses over the implications of this far too quickly. Traditional technical means did not appear nakedly, but only in the context of goals; people did not develop mere "means," but means in order to realize particular *ends*. A "neutral" means is evidently one divorced from this context—which Borgmann sees as a radically new phenomenon, as will become clear later in this chapter.

Pluralism developed as a response to the shortcomings of both substantivism and instrumentalism. It sees technology as many faceted—not just an autonomous power or a neutral means. Technologies are the outcomes of complex processes of evolution and interaction in which a play of myriad forces and influences determines which technologies ultimately arise and the forms that they take. Borgmann would count Latour among the pluralists. But Borgmann sees technologies as taking shape in a less complex process than do the pluralists. Technologies have radically altered the look of the world in a relatively short time, which "is possible only on the basis of strong and pervasive social agreements and by way of highly disciplined and coordinated efforts" (11). While the origins of specific technologies deserve to be explained by processes of interaction, Borgmann thinks that we should not ignore the presence of larger patterns at work no matter how complex and many-sided these individual technologies may be.

Borgmann is thus interested in developing a more adequate response to the tension between substantivism and instrumentalism than pluralism provides. He calls his approach "paradigmatic," insofar as he understands technology in terms of paradigms. Technology, in his view, contributes "a characteristic and constraining pattern to the entire fabric of our lives," a "pattern" or "paradigm" that "inheres in the dominant way in which we in the modern era have been taking up with the world" (3). He calls this pattern the "device paradigm." Borgmann's entire philosophy of technology consists of an attempt to bring to light and understand this paradigm inhering in the way in which human beings engage the world.

Borgmann therefore approaches technology in terms of specific technological devices, making his approach especially relevant to my concern in this book to effect a "turn toward things." Borgmann seeks the paradigm he is after not in the conditions of possibility of technology as did Heidegger and Jaspers, but in concrete objects. "Technology becomes most concrete and evident in (technological) devices, in objects such as television sets, central heating plants, automobiles, and the like. Devices therefore represent clear and accessible cases of the pattern or paradigm of modern technology" (3).

Although the technological paradigm is deep and far reaching—Borgmann says that "the pattern of technology is fundamental to the shape that the world has assumed over the last three or so centuries" (35)—it has gone nearly unnoticed. This becomes understandable when one considers the period in which this paradigm emerged, during the Enlightenment. The promising character that technology had at that time—and still has—served both to enhance and to conceal its enormous transformative power. This promise, as it was formulated, for instance, by Bacon and Descartes, consisted in both liberation and enrichment. The introduction of technology was not due to a simple desire to "dominate nature," as has often been claimed, but was done "with the aim of liberating humanity from disease, hunger, and toil, and of enriching life with learning, art, and athletics" (36). This promise has been partly fulfilled, and is still very much alive; technology is still viewed as the key to health and welfare. Because of the vagueness of this promise, which makes it easy to believe even though it cannot be exactly indicated what it entails, technology has been able to develop ever further. In this way our world has come to have an ever more technological character—with our attention focused on what technology promises rather than on the social changes it has brought in its wake. Thus the technological pattern has slipped unnoticed into modern existence, giving rise to the radically new "character of contemporary life."

Devices and Things

To sketch out what this "technological pattern" looks like in modern life, Borgmann begins by analyzing the relation humans have with technological devices. According to him, devices are the entities that fulfill the technological promise of liberation and enrichment. They do so by delivering *availability*. Technologies liberate human beings from needs and burdens and enrich their lives by making things that are difficult to acquire or realize more available with their help. Something is available, according to Borgmann, "if it has been rendered instantaneous, ubiquitous, safe, and easy"; "goods that are available to us enrich our lives and, if they are technologically available, they do so without imposing burdens on us" (41).

By way of example Borgmann analyzes warmth and its availability. In days of yore warmth was delivered to houses by a fire that burned in the stove or fireplace. This warmth was not instantaneous, ubiquitous, safe, or easy to obtain. Not instantaneous, because much work was required to obtain it: the wood had to be gathered or chopped up, or the coal bought, put in the coal shed, and fed slowly to the fire. Not ubiquitous, because not all the rooms in the house could be warmed in this manner. Not completely safe, because you could hurt yourself chopping the wood or burn yourself on the stove, and houses occasionally caught on fire. Not easy to obtain, finally, because work, skill, and attention were always needed to keep stoking the stove or fire. Technologies, however, have made warmth *available.* Today, someone who wants to warm his or her house simply turns up the dial of the thermostat on a central heating system—instantaneously, whenever she or he wants, and without assuming any risks or having to think about the matter further.

Availability is realized by what Borgmann calls "devices." Devices provide humans with what they had to obtain, in pretechnological times, with the aid of *things.* To stay with the example of warmth, what transpired to make warmth available was that a thing, the fireplace, was replaced by a device, the central heating system. A device, however, differs radically from a thing. A thing cannot be separated from its context or its world nor can it be divorced from our involvement with it: dealing with a thing requires us to engage with it and its environment. A device, on the other hand, puts out of play its context and does not require engagement; it does the work for us and without our involvement. The central heating system itself delivers all that is required to warm the house, needing nothing more from us than turning the thermostat—except when that, too, is automated. And the central heating system dispatches all the various forms of involvement that fireplaces and stoves require of us with themselves and with the world that arises around them—including the wood or coal merchant, the wood chopping or coal handling, and the necessity of gathering around the fireplace to keep warm. The central warming system does away with the need for all of these, and spares us the necessity of participating in the process by which warmth enters the house.

Borgmann calls what a device makes available a "commodity." Devices are in a position to deliver their commodity thanks to a second element in their constitution: their machinery. A central heating system relieves human beings of the trouble and effort that were needed to warm their houses by delegating the process to pieces of machinery: boilers, pipes, radiators, thermostats. Machinery makes up what Borgmann calls the "background of technology." It remains out of view as much as possible, because the handing over of work to machinery is precisely how a device makes it possible for humans not to have to do it. Were the machinery in the foreground, the device would require involvement instead of making commodities instantaneously available.

A device can thus be viewed as an entity that makes available a commodity on the basis of machinery that remains concealed as much as possible. Machinery has the status of a *pure* means, a means that is independent of the goal, divorced as much as possible from the commodities it delivers. Our ability to read the time, for instance, is not affected by whether the watch is powered by a spring or a battery. Technology divides things into means and ends and keeps these radically separate from each other.

The technological pattern, according to Borgmann, consists in the ongoing replacement of the presence of things by the availability of commodities delivered by devices. The inseparable bond that exists between nontechnological things and their context is delegated to a machinery that is hidden, so that we can enjoy commodities without having to engage ourselves with their production. In this way, the device paradigm divides things into commodities and machinery. The many-faceted role that the pretechnological thing played is now reduced to the single function by which the commodity is made available. The relation that human beings have with devices thereby becomes characterized by disengagement. Devices promote consumption, in Borgmann's use of the term; the consumption of commodities without engagement. The device paradigm is thus a pattern in which things that promote engagement become replaced by devices that invite only consumption.

While Latour describes a general structure to the way in which artifacts mediate human action, Borgmann uses his device paradigm to elaborate a specific consequence of that structure. He describes how the actions promoted by artifacts help to shape the involvements of human beings with their world. Things can be present in either an engaging way or in a nonengaging way. Borgmann's conclusion that only nontechnological things have the ability to engage human beings and that technological artifacts only invite disengaged consumption, however, is too hasty, as I shall shortly show. He focuses too narrowly on the forms of engagement that technology discourages or renders impossible, while ignoring that devices can indeed *promote* engagement as well.

TECHNOLOGY AND THE GOOD LIFE

Ultimately, Borgmann intends his analysis of the way in which devices shape human involvements with the world to illuminate how human existence is altered by technology. Technology, according to Borgmann, has developed into "a definite style of life" (92), which can be labeled as "consumption." This style takes shape not only by the disengaged ways of dealing with the world that devices promote, but also by the central role that technology plays in Western liberal-democratic societies.

Consumption

Borgmann's analysis of the technological mediation of existence is somber—in fact, too somber. According to him, technology's claim that it redeems its promise of delivering from want and enriching human life by fostering availability ignores any consideration of what true human enrichment consists in. Technology seeks to enrich human life by providing commodities, but it is difficult to see how the consumptive relation to commodities contributes to a meaningful human existence. The consumptive way of taking up with reality that technology invites cuts human existence off from its material and social contexts. The engagement with things and with other human beings that belonged inseparably to the involvement with pretechnological things is growing less and less. Borgmann names this kind of relation that human beings have to technology "implication." Humans are not dominated by technology, as the substantivists would have it, and neither do they treat it as a mere means. They are implicated in it, insofar as technology contributes a pattern to the way human beings deal with the world, a pattern that they take for granted in orienting themselves in the world. Humans do not choose technology; rather, technology forms the background for their choices. The pattern itself—the replacement of things by devices and of engagement by consumption—does not show itself, for it remains in the background to structure the way in which human beings deal with their world.

The most representative embodiment of consumptive existence, for Borgmann, is the so-called couch potato, who spends hours each day watching television—generally entertainment—and is deprived of the engagement with other persons and things that would occur by engaging in a conversation, reading a book, or playing a game, for example. American social scientists who study the ways people spend their time have discovered that the time devoted per day to political activities, attending lectures, reading, sports, outdoor activities, performing and listening to music, engaging in arts and crafts, visiting museums, and volunteer work—all of this time totaled up—is less than a quarter of the time that they devote to watching television (128–29).[3] Dutch researchers, too, have established a "dwindling number of filmgoers . . . declining readership of newspapers, books, magazines" and the "vanishing of parlor games in households" (Baudet 1986, 44).

Borgmann describes the emergence of this consumptive existence as the irony of technology: it promises enrichment but delivers impoverishment. Its disburdening character does not result in a way of living in which human beings experience

3. Borgmann draws here on Robinson (1977), among other studies.

engagement with their world. Initially, technological developments did "protect us from hunger, cold, disease, darkness, confinement, and exertion," and nobody would be willing to give these things up. But many new technologies can hardly be characterized in these terms: they assure us "that it is an imposition to have to open a garage door, walk behind a lawn mower, or wait twenty minutes for a frozen dinner to be ready" (Borgmann 1984, 140). The feeling of enthusiasm connected to technological innovations like these is "entirely parasitic":

> It is not animated by the full-bodied exercise of skill, gained through discipline and renewed through intimate commerce with the world. On the contrary, our contact with reality has been attenuated to the pushing of buttons and the turning of handles. The results are guaranteed by a machinery that is not of our design and often beyond our understanding. Hence the feelings of liberation and enrichment quickly fade; the new devices lose their glamour and meld into the inconspicuous periphery of normalcy; boredom replaces exhilaration. (140)

Though the great technological breakthroughs of the past have liberated human beings from misery, most technological innovation nowadays only serves to diminish our engagement with the world. That, according to Borgmann, is the irony of technology—it fulfills its promise of enrichment and disburdening in such a way that the disburdening it offers stands in the way of true enrichment.

Borgmann's analysis of the relationship between technology and consumption is open to criticism, though. His diagnosis is strongly reminiscent of Jaspers, who decried the emergence of "mass existence" in which human beings only consume mass-manufactured products and have no true ties with the world. The development of technology has indeed promoted a system of mass production and mass consumption, but it is rash to conclude that mass consumption leads to a pattern of human existence whose exemplar is the couch potato. Before developing this criticism further, however, let me first turn my attention to the way in which, according to Borgmann, human existence is also shaped by the technological mediation of its social context.

Liberal Democracy and the Good Life

Borgmann shows not only how devices shape the existence of the human beings who deal with them, but also, though more indirectly, how they shape the context in which that existence unfolds. Parallel to the way in which the hermeneutical perspective sees technologies as coshaping the cultural frameworks of interpretation that give shape to the experiences and interpretations of human beings, so the existential perspective sees technologies as coshaping the environment in which the actions and involvements of human beings unfold. In contemporary Western societies this environment is primarily provided by liberal democracy. According to Borgmann, liberal democracy and technology together shape the context in which human beings are invited to organize their existence according to the model provided by the device paradigm.

Characteristic of the liberal-democratic vision of society, according to Borgmann, is a particular way of bringing together the notions of freedom, equality, and self-realization. Human beings are considered equal in their ability to realize themselves, in such a way that they should be able to decide freely how to go about it. Equality is fleshed out as a matter of the opportunities for realizing oneself, freedom as a matter of free choice as to how to do so. Technology hooks up seamlessly with this specific constellation of ideas about freedom, equality, and self-realization. By making ever more goods available, technology makes it possible for human beings to realize their desires without imposing a content on how they go about it. Thanks to the surplus of articles of consumption that technology creates, human beings are able at last to make their own choices.

In this way, technology allows the practical realization of liberal-democratic ideals. It would be insufficient merely to strive formally for the equal opportunity for self-realization if human beings were not given the practical means for so doing. At the same time, these means cannot be based on a blueprint for "the good life," for the liberal democratic ideal is that the nature of the good life must be the responsibility of the individual alone. Technology makes it possible to steer between these reefs and gives substantial material content to the ideals of liberal democracy, while at the same time—apparently, at least—leaving open the question of the nature of the good life. By creating availability, human beings can both choose their own means for realizing themselves and make free choices from the available consumable goods. "Liberal democracy is enacted as technology," Borgmann writes (92), citing Daniel Boorstin approvingly when the latter asks us to "consider democracy not just as a political system, but as a set of institutions which do aim to make everything available to everybody" (88–89).

The form of self-realization that arises in this way, according to Borgmann, is not as neutral as liberal democracy would wish. Self realization, that is, assumes a technological shape. Though the intention of liberal democracy was ostensibly to leave the answer to the question of the nature of the good life up to human beings, a definite, associated answer has already implicitly been provided. While technology makes it possible to realize the stated aims of liberal democracy, according to Borgmann it also shapes the daily lives of human beings in accordance with the device paradigm. However all-encompassing the pluralism of liberal democrats may be in wishing to foster harmony and mutual respect between individuals with different perspectives of the good life, true pluralism, therefore, is nowhere in sight. The question of the nature of the good life has already been given an answer that is the same for nearly everyone: the liberal ideal of free self-realization appears in practice to involve mass consumption and work in order to make more consumption possible.

The "equal opportunities" that technology offers human beings, according to Borgmann, are therefore also far from neutral. Equality of opportunity is created by general welfare provisions such as an economic infrastructure, transportation systems, means of communication, and so forth, which are accessible to all. But these technological provisions are not neutral means, but actively shape humans' daily lives. "Surely they do not tell people when or where to travel, whom to call, or what to watch. On the other hand, a mountain valley that has been split by a road is no longer a place for solitary hiking. A perfect telephone system would suffocate the art of correspondence. And television at the least discourages municipal theaters and symphonies. We can see here the outlines of the technological pattern. Basic social and economic structures can be indifferent only as to the choice of commodities; but they are far from neutral as to the choice of engagement with things versus consumption of commodities" (96–97).

Borgmann's negative judgment of the impact that technology has on the shape of contemporary life is here once again evident. In the examples just mentioned of automobiles, telephones, and television he sees only threats to humans' engagement with their world. But this conclusion is too hasty. Before reaching such grand conclusions, it is advisable first to investigate more closely the contributions that technologies make to everyday human life. For on the basis of what has become clear by now it cannot be assumed at the outset that technologies merely impoverish the engagement that human beings have with their surroundings; automobiles, telephones, and televisions can just as well enhance specific forms of engagement. Automobiles allow people to meet more frequently with others who live far away, while telephones allow them to converse more often with each other. And thanks to television, for instance, millions of people were engaged with the disastrous explosion at a fireworks factory in Enschede that took place while I was writing this book, and thereby offered help to the victims. Later I will elaborate the positive contributions that Borgmann's analysis can make to the postphenomenological analysis of technological mediation, but first I shall present my objections to his diagnosis.

Just as little as liberal democracy, according to Borgmann, succeeded in keeping open the nature of the good life, so it was minimally able to realize its aim of creating equal opportunity. Huge disparities in personal income persist in liberal democracies, meaning that the poor are markedly less able to realize themselves than the rich. Yet up to now this has not posed any danger to the stability of our political system. Borgmann uses precisely this point to illustrate how strongly liberal democracy relies on technology. Belief in the promise that technology is the path to deliverance from want and provision of human enrichment by generating availability provides a kind of social and political stability. Since its onset, technology has brought economic growth and made available ever more goods. In consequence, people in the lower and middle classes acquire the perspective that, tomorrow, they will wake up to have what the rich have today (112). Far from promoting social unrest, the social inequalities present in contemporary society serve only to plant the device paradigm even more firmly in the saddle. "The peculiar conjunction of technology and inequality that we find in the industrially advanced Western democracies results in an equilibrium that can be maintained only as long as technology advances" (113).

So long as technology provides the pulse and the stabilizing factor of democracy, according to Borgmann, politics can never become a place where the seminal questions about our social life, including the nature of the good life, can be meaningfully debated. Politics is all about the sharing of commodities. It has itself become a device that engages ever fewer people. Borgmann calls politics a "metadevice of the technological order," and notes that "The calls for participatory democracy which are oblivious to the substance of politics and merely recommend new forms of transaction are pointless and will remain inconsequential. One may as well call for participation in pocket calculators" (113).

Focal Things and Practices

If we ever become able to pose again explicitly the question of the good life, Borgmann says, we will find that alternatives to technological consumerism do exist. These alternatives, according to Borgmann, cannot consist in radically rejecting technology, but rather in reforming it. The reforms that he proposes can themselves be realized only with the device paradigm. Borgmann does not want technology to be "enclosed in boundaries" but to be "related to a center"

(168). That center would be provided by what he calls "focal things and practices." Focal things for Borgmann are things that draw together human involvements, things that invite engagement with themselves and what they make possible. They are "matters of ultimate concern that are other and greater than ourselves" (169). In addition, they "are concrete, tangible, and deep, admitting of no functional equivalents; they have a tradition, structure, and rhythm of their own. They are unprocurable and finally beyond our control" (219). Moreover, focal things have a "commanding presence" and "centering power" in our dedicated dealings with them (Borgmann 1992a, 119–20).

Focal things promote what Borgmann calls "focal practices," by which he means ways of dealing with the world that are characterized by engagement. Only in focal practices are focal things present as such. "A focal practice, generally, is the resolute and regular dedication to a focal thing," and, conversely, a focal thing "sponsors discipline and skill which are exercised in a unity of achievement and enjoyment, of mind, body, and the world, of myself and others, and in a social union" (Borgmann 1984, 219). To speak about focal things and practices requires a specific discourse that Borgmann calls "deictic," a term that means something like "indicating": focal things and practices can be indicated or pointed at, but their existence can never be proved or analyzed with scientific exactitude. This does not mean, though, that deictic discourse is arbitrary or insignificant; on the contrary, it consists in sincere descriptions of the value a particular practice centered around a thing has for a person.

In *Technology and the Character of Contemporary Life,* Borgmann discusses two concrete examples of focal practices: running and the culture of the table, with the associated focal things being the running shoes along with the environment in which one runs, and the kitchen utensils along with the food that they are used to prepare. Long-distance running—or trekking—is an intensive way of being present in one's surroundings. A run or trek is physically demanding, but it provides an experience of freedom and pleasure, and a special kind of involvement with the surroundings along the route. The culture of the table—what is involved in preparing, sharing, and enjoying a meal—requires considerably more effort than warming up something in a microwave. One has to attend to the food: contemplating what to prepare, purchasing and washing the ingredients, carefully cooking it, and eating it attentively and with enjoyment. Both practices are valuable in and of themselves, and are not pure means of "getting from one place to another" or of "obtaining nutrition." They involve not availability but engagement.

Reforming technology, according to Borgmann, will involve making room for such focal things and practices. The technological pattern tends to crowd out the kind of presence that these require. Because focal practices are preeminently engaged ways of dealing with the world, they are in a position to break the technological pattern. Borgmann says explicitly that his reform of technology does not entail a repudiation of it. Technology is merely assigned a different status, and becomes a condition of this reformation instead of an end in itself. The device paradigm itself must be used as a means—for the end of fostering focal practices. In proposing that we break the technological pattern, Borgmann is not calling for us to retreat to pretechnological enclaves, but rather to keep technology more at bay, more at the periphery of our lives. And it is precisely the ability of technology to promote availability that makes it preeminently useful to provide room for focal things and practices. "Technological aids to human life have become indispensable, an no one should wish them away. The indispensability of some sort of technological machinery is no misfortune. It is compatible with eloquent reality, as are many of the technological commodities" (Borgmann 1992a, 119). To make such a reformation of technology possible, according to Borgmann, requires the emergence of a "public deictic discourse," able to address not just standards but also quality of living—not just affluence, but also wealth, in Borgmann's words. Political debates cannot be limited to the sharing of commodities, but also have to range over the quality of life.

BEYOND ALIENATION

Borgmann's concepts enable the existential dimension of technological mediation to be fleshed out further, but a few critical remarks should be borne in mind. His diagnosis comes dangerously close to the alienation thesis that was criticized in the first part of this book. The power of Borgmann's analysis consists in his approach of technology in terms of specific artifacts rather than reducing it to its conditions of possibility, as did the classical philosophy of technology. The pattern that he perceives as organizing our existence is not something a priori of which technology is the concrete realization; Borgmann derives it by studying the ways in which devices invite particular ways of dealing with them and with the context in which they function—he even speaks of "the power of products . . . to shape our conduct profoundly" (110). However, the particular content of this pattern—that the involvements that human beings have with the world are diminishing more and more due to devices—is questionable.

Borgmann's claim that technology leads to an impoverished, consumptive existence strongly recalls Jaspers's philosophy of technology. I have already criticized their somber diagnoses using simple counterexamples, but this intuitive critique deserves to be worked out more systematically. I will first point out that Borgmann's concept of engagement undergoes a shift in meaning as his argument develops, making the alleged impoverishment of modern life at least ambiguous. Then I will show that the exclusive alternative he offers between engagement with nontechnological things and disengaged consumption of technological devices is untenable. The pattern that he outlines does not do justice to the role of technology in human existence, however enlightening it is to describe it in terms of involvements.

Two Forms of Engagement

However sympathetic we may be to many aspects of Borgmann's account, there is nevertheless a snake in the grass. He claims that devices diminish the engagement of human beings with their environment and invite consumption. Instead of stoking the fireplaces today people warm their houses by turning a dial on their thermostats; instead of cooking food themselves they warm up prepared food in microwaves; instead of traveling by foot or horseback they take the train or drive. The engagement that is thereby lost, according to Borgmann, can be recovered if people devote themselves to focal things and practices. But when the engagement that technology causes to be lost is compared with the engagement brought by focal practices, a striking difference emerges between these two kinds of engagement.

The engagement with things that is lost in the consumption of commodities consists of the effort and difficulty that was traditionally required to obtain something. The type of engagement Borgmann sees as compensating for this loss is supposed to be regained in focal practices, which are valuable in themselves without serving a particular end, and which constitute meaningfulness. The focal things in these practices represent a completely different class of objects than pretechnological things, while focal engagement is an entirely different kind of involvement than the effort and exertion from which technology delivers us. Nontechnological things call for engagement in the sense that people have to apply and exert themselves in order to use them. A thing is never a pure means, but calls attention to itself and to its environment. In contrast to a technological device, a thing does not provide the most convenient path to achieve a goal, but involves its users in the realization of it. The engagement that focal things call for is of a completely different sort. Borgmann does not describe it in terms of effort and exertion, but of meaningfulness. And he speaks about focal things and practices in terms of "orienting one's life" and "realizing one's aspirations"; focal things, he says, are "eloquent reality" and possess a "commanding presence."

No doubt some dimension of meaning can be found in the encounter with pretechnological things, but it romanticizes the past to claim that the disappearance of things like drawing water and gathering wood involves a loss of meaningfulness. People did such tasks not because they were intrinsically valuable, but because of the ends that these tasks served—ends that technologies can realize in a different way. Faucets and central heating systems have made it unnecessary to draw water from the well and to gather firewood, which has indeed helped to diminish human contact with the others one would meet at the well and in the forest. But focal practices do not serve any specific goal. They do require dedication and effort, but they are not "useful" the way that drawing water and gathering firewood are; rather, they are meaningful in themselves, as practices such as playing the piano or cooking and eating an elaborate meal with others. Focal practices include an existential involvement of human beings with their world, a manner of presence that Jaspers called "immersing oneself in the world" and that he, too, wanted to recover.

By distinguishing between practices that require effort and those that produce meaningfulness as two different forms of engagement, I do not mean to challenge the relevance or importance of thinking in terms of practices that give meaningfulness. What is problematic in Borgmann's analysis is not the dimension of meaningfulness, but the ambiguity in his concept of engagement. Stating that technology diminishes the efforts that people have to expend is quite another matter than stating that it diminishes the possibilities of experiencing meaning and leading an engaged life. But when Borgmann outlines the device paradigm, he gradually shifts from the first meaning to the second and ends up blending the two. Devices not only diminish the exertion that is required to get something, but at the same time discourage focal involvement, Borgmann holds. The person who watches television—Borgmann's emblematic example of a device paradigm—is not so much spared the trouble that he or she would be burdened with without one, but is rather invited to participate in a consumptive practice that takes up space and time that might instead be occupied by a focal practice.

Distinguishing between these two forms of engagement makes clear that, on the basis of Borgmann's theory, technology primarily leads to a reduction of effort and only in an indirect way threatens focal engagement. By using a device instead of a thing, people can do what they want with less effort—but a device can never be a true alternative to a focal practice, first and foremost because such a practice is never straightforwardly aimed at the realization of an end for which technology could provide a more efficient means. People do not run marathons for transportation purposes, nor prepare festive meals to satisfy their hunger. Within the context of Borgmann's theory, therefore, technology cannot directly erode focal practices. Focal practices are threatened only by the "technological pattern," which invites people to be present as consumers in their world not directly, as a device does, but indirectly, by the device paradigm. If people give up focal practices, they do not do this because they use technological devices, but because they are entirely submerged in the consumptive attitude that the use of devices invites. It can be doubted, however, that such an attitude is actually this pervasive in human existence and that it can be derived from the actions that devices invite. While technologies, that is, in many cases indeed require less effort than pretechnological things, they are also often in a good position to make possible focal engagement.

Engaging Devices

Technologies generally diminish the amount of effort that is required to obtain goods, but their role in "focal engagement" is more ambivalent. Someone who drives a car does not have to run or bicycle; someone who uses a word processor does not have to retype the entire manuscript when revising; someone who uses a washing machine does not have to hand-wash and hand-wring the clothes. But in the case of focal engagements the situation is completely different. For every one of Borgmann's examples of devices that invite consumption one could cite many counterexamples.

An electronic piano, for instance, facilitates engagement as well as an acoustic one. To be sure, it saves us some amount of effort: it does not have to be tuned, is easier to move around, and can be played with headphones so as not to bother the neighbors. But even though the sound quality of electronic pianos (at least until recently) is poorer than that of most acoustic pianos, as is their response to subtleties in the attack (again, until recently), it does allow a user to be fully engaged with music. Many students are only able to practice thanks to such devices. Because of their frequent relocation or to prevent the noise from disturbing others, possessing an acoustic piano for them is not a real option.

The CD player, too—which Borgmann uses as an emblematic example of a consumptive technology—in most cases is more engaging than consumptive. Thanks to this device, people can involve themselves intensely with Bach's cello suites without going to the trouble to find a cellist and arrange a performance, or hear various interpretations of the "Well-Tempered Clavier" without having to attend concerts by all the great pianists. Without CD players people could not be as engaged with music as they are, for having to depend on live music—even though that has additional value above a recording—would make this engagement next to impossible. Technology unquestionably has made it possible for more people to listen to more music. CD players, along with radios, record and cassette players, and television, have made music less elite, and currently it is one of the most enjoyed of all art forms. Complaints about cultural decline among the younger generation caused by the technological availability of music have it backward; young people make up a huge group of music lovers, for whom even special radio and television programs are made. And for them, music is generally more than a device to provide a nice background ambience, but rather something that has a value in itself. The ability of technology to strengthen rather than weaken involvement with music was already clear at the time of the introduction of the phonograph, which gave rise to ongoing "associations between record lovers and magazines about records, record reviews, technical information, and other sections" (Baudet 1986, 64).

Even television, Borgmann's textbook example of a technology that invites consumption, can engage people. Not so much with its machinery—except perhaps for the hobbyist—but by the content of some of its programs. Not every program aims at mere diversion and entertainment. News programs can make people aware of parts of the world affected by war and hunger, expanding the potential scope of their engagement and of their charitable contributions. Arts programming can spread awareness of contemporary cultural movements and developments, and television has made movies accessible to larger audiences. Moreover, the idea that television viewing is purely a matter of individual consumption is simply wrong, as is evident to anyone who has ever watched a live broadcast of a tense game of any kind together with other people. And those who found parlor games engaging surely would have little objection to most entertainment programming, even though these seem to be the first to deserve the predicate "consumptive." Many game shows are scheduled in prime time, and thus can foster a family's ties by allowing its members to play along, compete with each other, and in some cases even to participate via telephone or mail (cf. De Meyer 1994, 65–67). I do not see the difference between this and an evening of playing Monopoly or Sorry.

In some cases technologies even enhance engagement in the sense of effort. In the usual version of the story of the microwave, the moral is that it saves a lot of work and stimulates a consumptive comportment toward food, turning meal preparation into a comparative snap. Though there is certainly a grain of truth in this, as I argued in the introduction, research by Cockburn and Ormrod tells a somewhat different story in which the labor saving does not always happen (cf. Cockburn and Ormrod 1993). When the microwave is not used for frozen foods and is used in ordinary cooking, its speed is used in the service of demanding eaters or those with special needs. While before a single meal was prepared for everyone, with devices like the microwave it is now possible to prepare meals individually, for instance with different seasonings for everyone (143). It is reminiscent of the law of conservation of travel time: just as the emergence of faster means of transportation has not had the effect of reducing commuting time but rather lengthening the distance over which people commute, so faster cooking equipment has not shortened meal preparation time but instead invited a more complex cooking process.

These counterexamples show that Borgmann selectively supports his outline of the technological pattern. He perceives that particular pattern only by virtue of the fact that he has an incomplete picture of the role of technology in involvement. Borgmann does not see that technology can not only reduce engagement but also amplify it. Technology gives rise not only to disengaged consumption, but also to new possibilities for engagement. This conclusion also raises questions about Borgmann's use of the concept of consumption to describe a disengaged way of dealing with reality. It appears that consumption can be better understood as the making use of products in which the amplification and reduction of engagement are entwined together in an ambivalent way. Borgmann only admits a single aspect—disengagement—of the implications of technology for the involvement of human beings with their world, and systematically disregards the ability of technology to invite new forms of engagement.

These new forms of engagement do not always happen, of course, but neither does technology always involve disengaged consumption. Nevertheless, the amplification of engagement is just as surely a property of technology as its reduction. The fact that something came to pass via a technological process does not warrant the conclusion that it is a disengaged article of consumption. The emergence of Bach's cello suites from a CD player does not imply that no engagement is possible; the fact that someone went someplace by airplane is not evidence that that person is not intensively present. Technology indeed makes things available, but the lack of human involvement in the process does not mean that humans are not involved in the product. Reduction of one form of involvement usually goes hand in hand with the amplification of other forms.

These considerations remove the barb of the alienation thesis from Borgmann's theory. His claim that technology runs contrary to engagement and invites a disengaged and consumptive existence is untenable. On the existential level as well as on the hermeneutical one, technologies can be seen to play an ambivalent role in shaping the human-world relation. Borgmann's outline of the device paradigm only does justice to a limited portion of the role that technologies play in human existence. At the same time, his approach to technology in terms of engagement provides a perspective that is relevant to the postphenomenological analysis of the role that technology plays in human-world relations. Involvement, in the form both of effort and meaning-giving engagement, needs to be treated as a dimension of technological mediation and not as something that technology excludes or renders impossible.

MEDIATED ENGAGEMENT

The preceding critique of the problematic features of Borgmann's approach opens the way for a translation of his analysis into the postphenomenological perspective. To do so, I shall describe Borgmann's account of how technologies shape involvement in terms of mediation. This will put in place the final step in the analysis of the existential dimension of technological mediation. The previous chapter sought, with the help of Latour's concepts, to illuminate the structure of the technological mediation of action. Borgmann's concepts allow us to analyze how human engagement with reality, which results from these actions, is technologically mediated as well, and thus how technology affects how human existence takes shape.

Devices and Involvement

The mediation of human-world involvements, as I discussed earlier, has a structure of amplification and reduction. In just the way that technological artifacts mediate perception by excluding certain interpretations of reality and promoting others, so can they make possible certain kinds of actions and inhibit others. But instead of speaking about the reduction of engagement, as Borgmann does, it is preferable to appropriate Latour's terms and speak about its translation.

As in the case of the transformation of interpretations, the translation of involvements takes place in two ways, one direct and one more indirect. The direct way of translation concerns the mediation of action. By encouraging particular actions (invitation) and discouraging others (inhibition), some forms of involvement are called forth and others suppressed or excluded. The use of a

CD player discourages making music oneself, but invites listening to music. But in the same way that technologies mediate human interpretations not only directly in perception but also indirectly by affecting interpretive frameworks, so they mediate human involvements indirectly via the social context in which human existence plays itself out. The social context for our existence, according to Borgmann, is shaped by liberal democracy, which defines "the good life" in terms of the consumption of mass-manufactured products, which technology has made available in surplus quantities.

The pattern of mass consumption shapes the existence of human beings; it provides the opportunity for products to mediate human actions and thus their involvements. This mediation, again, must not be thought of only in terms of reduction. While in some cases mass consumption diminishes human engagement, it can also open up new forms of engagement, new existential space in which human beings can realize their existence.

In analyzing the technological mediation of involvement—which includes both effort and (meaning-giving or focal) engagement—it is worth the trouble to determine to which aspect of reality the involvement directs itself. In this chapter I have implicitly pointed out three different variants of involvements: they can concern the artifact itself (playing a piano requires effort), its environment (a fireplace requires that wood be gathered and chopped up), or the product that the artifact makes available (the CD player makes it possible to enjoy music).

Involvement with the artifact itself is often diminished by technological devices, for their physical presence (machinery) is usually subordinated to their function (delivering commodities)—they are meant to call as little attention to themselves as possible to reduce the effort needed to use them. But devices can also invite involvement, both in the focal sense (an electronic piano, for instance) and in the sense of effort (dental equipment, for instance, which must be constantly cleaned and maintained).

The same is true in the case of involvements with the environment of artifacts, for these can be diminished by technological devices, but they can also call forth new forms of involvement. The multitude of activities in which people have to participate in connection with nontechnological things in order to get something from them, such as collecting wood in order to stoke the stove and walking to the well in order to draw water, are often rendered unnecessary by the use of devices. But then there are also devices that make practices possible in which we could indeed speak of increased involvement with the surroundings, such as the effort expended by the gardener who at the moment I write this is cutting the lawn beneath my window with a power mower, or the focal engagement of a cook who is using an electric blender to concoct a superb soup.

The ambivalent character of technological mediation is most explicitly evident in the third variety of involvement, which is directed at that which the devices make available: their "commodities," in Borgmann's words. Even though many technologies do discourage involvement with themselves and with their environment, in many cases they nevertheless do encourage involvement in what they make available. Humans can be focally engaged with the music that comes out of a CD player, for instance, and word processors allow them to be intensely occupied with the ready-made formatted text on the screen.

Involvement and Human-Artifact Relations

Thus far, I have analyzed the technological mediation of human behavior and involvements as the mirror image of human perception and interpretations that were central in the hermeneutic perspective. As we saw in Chapter 4, human experience—and thus the way reality is present to human beings—is transformed in the technological mediation of perception and of cultural frameworks of interpretation. Analogously, we now see that human existence is translated—and thus the way in which people are present in their world and the way their involvements take shape—by the technological mediation of human action and its social context. In this existential perspective on mediation, however, one aspect still need to be developed, one that Ihde has elaborated in his hermeneutical perspective; the nature of the different kinds of relations between human beings and artifacts, from which these artifacts can play their mediating roles.

On the basis of Ihde's work, a tripartite distinction can be made between types of human-artifact relations, as we saw: relations of mediation (human–technology–world), alterity relations (human–technology/world), and background relations (human/technology–world). Perception, unsurprisingly, appeared to be mediated principally by relations of mediation, in which a distinction is drawn between the embodiment relation and the hermeneutical relation. This mediated relation also plays a key role in the mediation of action, but alterity relations are important here as well. Of both types of relations of mediation, only embodiment relations are relevant in the analysis of the mediation of action. On the level of the mediation of action, the hermeneutical relation only concerns phenomena such as virtual reality, for in them action plays itself out in a representation of reality rather than in reality itself.

Embodiment relations, however, have a remarkable content here. The way in which artifacts are present in this relation can be indicated as "readiness-to-hand," as we saw in Heidegger's tool analysis, which appeared to be extendable to technological equipment and devices in general—automobiles and power drills, for example, can also be ready-to-hand. Borgmann's work, however, allows us to see that artifacts can be ready-to-hand in two different ways—in an engaged or a disengaged way. This connection between readiness-to-hand and engagement is remarkable. For Heidegger, an essential characteristic of a ready-to-hand artifact was that it does not call attention to itself, but withdraws from attention in order to direct attention to the work to be done with it. Borgmann himself formulates a related thought when he claims that the machinery of devices places itself as much as possible in the background, so that human beings only have to attend to what the device *does* for them. If the object did not withdraw, runs Heidegger's and Borgmann's thinking, it would be impossible to use it; it would not be functional if it called attention to itself. But the existence of "engaging devices" calls for a nuancing of these analyses.

Some artifacts, such as a piano, indeed create involvement in their functioning, and thus give rise to the intriguing situation of both withdrawing from people's attention and calling attention to themselves at the same time. Someone who plays the piano is directed toward the music and at the same time is substantially involved with the piano itself. When the same piece is played on a CD player, the artifact that mediates between the person and the music is present in an entirely different way. The machinery of a CD player disappears into the background, withdrawing so that people are only engaged with the music and not with its means of production. A piano, however, is never entirely ready-to-hand, but neither is it exclusively present-at-hand—its machinery is not completely in the background, but not entirely in the foreground either. Readiness-to-hand and presence-at-hand, therefore, cannot be conceived as two modes of human-artifact relations, but rather as the termini of a continuum on which this relation unfolds.

To make things more complex, not only Heidegger's binary opposition between readiness-to-hand and presence-at-hand needs to be challenged, but also the idea that artifacts need to be ready-to-hand in order to be usable. Artifacts mediate human action and involvement not only via embodiment relations but also via alterity relations. Borgmann's fireplace example illustrates this nicely. A fireplace is not "embodied" or ready-to-hand when it is used. It has an entirely different position in human-world relations than, for instance, a pair of glasses. A fireplace does mediate these relations, but from a present-at-hand position instead of withdrawing into readiness-to-hand. It remains present and demands involvement with itself, and by demanding these involvements it mediates human action: it needs to be cleaned, stoked, filled, it requires wood to be gathered and split, and humans have to sit around it in order to enjoy its warmth. From a present-at-hand position (or alterity relation), a fireplace helps to shape the involvement of human beings both with the fireplace itself and with its environment.

To complete the picture, in the involvement that takes place in alterity relations a distinction can be made between "effort" and "focal engagement," and the involvement can be with the device itself, its environment, or with the products that it makes available. This makes for a multiplicity of ways in which artifacts mediate human-world relations. An automobile enthusiast can have a focal engagement with classic cars, while I direct my efforts to my bicycle only when it breaks down. A sewing machine allows fashion designers to be focally involved with its environment (fabric, models, fashion-shows, and so forth), while an automobile has an entirely different environment that requires all kinds of efforts: gas stations, garages, highways, a bucket of suds, the things you can deliver to friends who do not own cars, and so forth. An airplane makes far-off lands available, but people can still be engaged when they get there; a copy machine makes lots of printed text readily available, which may bring with it much in the way of bureaucratic efforts.

CONCLUSION: THE MEDIATION OF ACTION AND EXPERIENCE

A Postphenomenological Vocabulary

In the preceding chapters I have tried, by means of a critical review of the positions of Ihde, Latour, and Borgmann, to develop a postphenomenological vocabulary for analyzing the mediating role of artifacts. From a hermeneutical perspective, artifacts mediate human experience by transforming perceptions and interpretive frameworks, helping to shape to the way in which human beings encounter reality. The structure of this kind of mediation involves amplification and reduction; some interpretive possibilities are strengthened while others are weakened. From an existential perspective, artifacts mediate human existence by giving concrete shape to their behavior and the social context of their existence. This kind of mediation can be described in terms of translation, whose structure involves invitation and inhibition; some forms of involvement are fostered while others are discouraged. Both kinds of mediation, taken together, describe how artifacts help shape how humans can be present in the world and how the world can be present for them. The accompanying table provides the most important concepts from each perspective.

TABLE I
A Postphenomenological Vocabulary

Hermeneutic	***Existential***
Experience	Existence
How reality appears to humans	How humans appear in their world
Perception (microperception)	Action
Interpretation	Involvement (efforts and focal engagement): —with the artifacts themselves —with the contexts of artifacts —with what artifacts make available
Transformation	Translation
Amplification	Invitation
Reduction	Inhibition
Constitution of objectivity	Constitution of subjectivity
	Most Relevant Human-Artifact Relations
—embodiment relations	—embodiment relations
—hermeneutic relations	—alterity relations
	Points of Departure
—artifacts mediate perception and context of interpretation (macroperception)	—artifacts mediate action and context of existence
—experience takes shape as perception interpreted within a context of meaning	—existence takes shape as action involved in a context of existence

The postphenomenological perspective offers a completely different view of technology than the classical philosophy of technology. In contrast to the latter's fear that technology alienates us from reality and from our authentic existential possibilities, the postphenomenological perspective offers a rich and variegated picture of technology, which does justice to its ambivalent status. That picture can only arise if technology is approached in terms of artifacts, rather than as the expression of a calculative and manipulative way of thinking or as a functional appendage of society. When this happens, the implications of technology for human existence and meaning appear to be far reaching—technology codetermines both human subjectivity and the objectivity of the world. But at the same time these implications emerge as more subtle and nuanced than the manner in which the classical philosophy of technology revealed them. While the classical philosophy of technology speaks of technology as involving a loss of meaning, postphenomenology speaks of transformations of the ways in which reality can be present for humans; and while the classical philosophy of technology speaks of the loss of self, postphenomenology speaks of translations of the ways in which human beings can be present in the world and realize their existence. In these transformations and translations, some ways of being present are strengthened and others weakened.

An Example: The Personal Digital Assistant

The value of this postphenomenological approach to technology does not lie merely in the fact that it provides a framework with which to critique the classical philosophy of technology, but additionally in the fact that it makes possible a more careful and thorough investigation of specific technologies. Its vocabulary makes it possible to describe technologies not simply in terms of their functionality but also as mediating the relation between human beings and their world. By way of illustration—and to show the interrelations between the key concepts of the postphenomenological vocabulary—consider the personal digital assistant (PDA), a device that seems destined to play an ever greater role in everyday human life.

PDAs are small, stripped-down versions of personal computers. They do word processing; have clocks, calendars, and address books; and even provide email and Internet access—all packaged in a small unit, able to fit in a jacket pocket, and weighing only a few ounces.

A PDA affects its users in more ways than just its via functionality, as I can attest from my own experience. For someone who rides the train often, for instance, it can substantially reduce the pressure of work, for on the way one can quietly work on an article or letter, send email, and more. But this possibility can also have the opposite effect—because it makes it possible to work virtually each moment of the day, it can rob one of relaxing train rides.

During a train ride, a PDA not only mediates its user's pressure to do work, but also translates the specific actions of its user, from things like "reading," "looking at the landscape," or "conversing with one's neighbors" to "working on one's text." A PDA invites writing and involvement with the text and discourages other activities such as paying attention to the surroundings, except when it leads curious fellow travelers to ask about it. The involvements the PDA makes possible, moreover, can involve both effort and focal engagement: when one works on a text with a PDA, this activity can be experienced as intrinsically valuable; when taking minutes at a gathering, the effort-related character of the activity is most likely to be predominant. The PDA also invites another kind of involvement: toward itself as a physical object. Because it is small, expensive, and fragile, it can easily be lost, stolen, or damaged. That encourages its owner to treat it like a wallet, something of whose location one is always conscious even when it is not being used.

These last mediations of involvement occur on the basis of alterity relations, for the human actions are not executed *through* the PDA but are directed explicitly *at* it. When one writes with a PDA, one embodies it, for the user is then directed to the text being written rather than to the device itself. The tiny (on-screen) keyboard from time to time tends to change the embodiment relations over into alterity relations; certainly while learning to use the PDA one is apt to make numerous typing mistakes. On the basis of its embodiment, the PDA mediates (existentially) how somebody is present in his or her environment—when somebody on the train, for instance, is writing instead of looking about, or involved with a text instead of a newspaper or fellow travelers—but also (hermeneutically) how the world is present for her or him. The world, that is, has become reduced to a text—and a rather limited kind of text at that, for the PDA's screen presents only a tiny part of it at a time, because of which it is advisable to work either on an almost finished version of a document or to compose a first draft. This transformation of the way the text is present, however, involves not only reduction but also amplification. A person who only has a printed version of a text available can only read it, whereas a PDA also allows editing.

The dimensions of action and perception, with the associated reductions and amplifications of interpretations and involvements, are also present when the PDA is used for emailing. The presence of the interlocutor is reduced to the text that he or she writes, but this text at the same time amounts to an amplification, for without email the interlocutor would not be present at all. And not only is *perception* of the other possible via email but so is *action:* one can reply. This translates the interaction when compared to a conversation. The other is not physically present, yet one can more carefully consider one's interventions.

A PDA, in short, helps to shape the way in which its user and the user's world are present to each other. Even on the train, a small computer can mediate one's entire world, directing one's attention to a text unscrolling in one's lap instead of the landscape outside. One can contact others via email, but this contact has a completely different character than conversations with a neighbor. And one is sure to handle the PDA carefully lest it be damaged, lost, or stolen. A PDA helps to shape its user's existence and experience; it shapes specific aspects of its user's subjectivity and the objectivity of that user's world. It is more than a functional instrument and far more than a mere product of "calculative thinking." It mediates the relation between humans and world, and thus coshapes their experience and existence.

PART III

7

Artifacts in Design

INTRODUCTION

In the previous chapters I have developed a postphenomenological vocabulary for analyzing the mediating role of artifacts in the relation between human beings and their world. With the help of this vocabulary, it is possible to describe technology in a more nuanced light than the terms of the classical philosophy of technology allowed. This approach does not reduce the role of technological artifacts in the lifeworld to a loss of meaningfulness and authentic existence, as did Heidegger and Jaspers. Technological artifacts indeed close off some possibilities by the way they mediate experience, but they also open up new ones. Technology mediates our behavior and our perception, and thereby actively shapes subjectivity and objectivity: the ways in which we are present in our world and the world is present to us.

This new vocabulary can also be applied outside the context of the philosophy of technology. Because it explicitly addresses the role of objects in their environments, this vocabulary offers the prospect of an application to the work of designers, who are continually creating artifacts embodying the kinds of mediating capacities described in this book. Here I shall restrict myself to industrial design, precisely because this discipline concerns artifacts that play a large role in everyday social life. Most contemporary approaches to industrial design pay little attention to the mediating role of artifacts. In this chapter I propose a broader approach that might be called a "material aesthetics." This approach aims to make a contribution to the ongoing discussion of environmentally sound industrial design, and thus to demonstrate the practical value of the postphenomenological perspective.

THE MATERIALITY OF THINGS

Industrial design generally treats products from one of two perspectives: their functionality and their sign-value. A product must first of all be functional; it must do what it was designed and manufactured to do. Besides this, it has meaning or sign-value: human beings are drawn to particular product styles and not to others, and use a product to express the lifestyle to which they (want to) belong. In order to obtain a sharper picture of current approaches to artifacts in industrial design, it is helpful to discuss this distinction between function and meaning and the role that distinction plays before investigating what light the postphenomenological perspective sheds on this issue.

Functions and Signs

In his book *Order and Meaning in Design,* the Dutch design theorist Wim Muller analyzed the functionality and meaning of products as two different aspects of their utility value (Muller 2001, 14). According to Muller, designers aim to join two forms of utility in the products they are designing: "material utility," indicating the practical usefulness of the product, and "social-cultural utility," consisting in its social and cultural meaning. A yacht, for example, is built not only in order to go sailing, but also to express the status of its owner. To these two aspects the two different roles of the designer correspond, though of course they are closely connected since they have to be brought together in the final product. The material utility is often considered the domain of the engineer, whereas the social-cultural utility is located in the domain of the designer (14).

But even within the work of the designer the distinction between function and meaning crops up, as I already indicated in the brief discussion in Chapter 1 of the history of design. There, I spoke about the transition from modernism, characterized by a functionalist approach to design with ornamentation kept to a minimum ("form follows function"), to postmodernism, which recognizes that meaning is central to products and promotes the development of myriad product styles addressed to different lifestyles ("form follows fun"). These two periods are flanked by transitional phases like the "Arts and Crafts Movement," "*Jugendstil*," "art deco," and "pop-design."[1] Here, the distinction between the functionality and the meaning of products is not bound up with that between the work of the engineer and that of the designer, but rather involves different approaches within design itself. Muller's distinction between material and social-cultural utility is also addressed by design theory via the distinction—originally made by Jochen Gros—between "practical functions" versus "product language functions," as figure 3 shows.[2] Practical functions concern the function for which the product was originally intended: sitting in a chair, riding in an automobile. Product language functions refer to the way products function as "signs" within "languages." They not only do something for human beings, but also have a signifying character: they indicate, refer to something. A chair and an automobile show something about the taste of their owners. They do that on the basis of their formal characteristics, such as the spatial features of their construction, indicated by designers as their formal-aesthetic functions. In connection with these formal-aesthetic functions products also have semiotic functions; their specific form gives them a specific meaning.

Two general classes of semiotic functions can be distinguished. A first is that products have "denotative functions": they denote for what and how they are to be used. Thus a chair refers to the possibility of sitting, at least in a context in which objects with that kind of shape are used for that kind of purpose. Denotative functions can also concern parts of the product as well as the whole. An object, for instance, can denote what its "up" and "down" sides are, how it is to be switched on, and so forth, thanks to such things as clearly visible and handy legs, knobs, and so on. Denotative functions "communicate an intended or expected kind of interaction" (299).

The second class of semiotic functions consists of "connotative functions." These are the kinds of functions that feature so prominently in postmodern design. Connotative functions turn the product into a symbol: they ensure that it represents the lifestyle with which its users identify, or want to identify. Products, for instance, can be seen as "solid," "sturdy," "traditional," "outgoing," "trendy," "wild," and so forth. Designers use surveys of and market research into various kinds of lifestyles in their work, in the form of such things as the so-called AIO

1. See Heskett (1980); Sparke (1986); and Ramakers (1986).
2. This figure is taken from Muller (2001, 299).

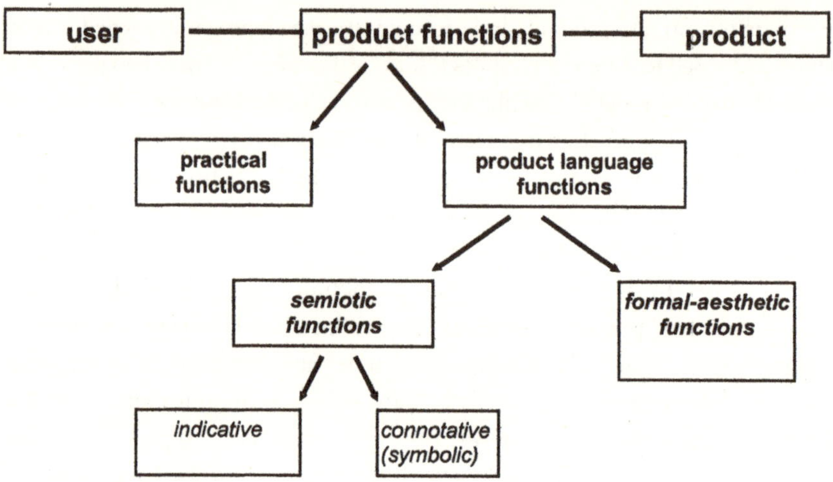

Fig. 3 Product functions

data—activities, interests, and opinions (Bürdek 1996, 231). These kinds of surveys classify groups as "experimenters," "followers," "competitors," "socially conscious," "integrators," and so forth (Muller 2001, 317).

Such symbolic functions determine the social-cultural utility of products, forming what Muller calls their "secondary function," along with their primary function or "material utility." In Muller's words: "For most of our utensils, we can distinguish secondary functions. This is true for our clothes, our houses, our furniture, and maybe especially for our cars. The statement that these were designed because of their primary function is difficult to maintain. In many cases, secondary functions are deliberately anticipated with even an increasing tendency to differentiate the secondary function features of a product toward the life style of specific target groups (consumer types)" (316).

This semiotic perspective, in which products are viewed as bearers of signs, plays an extremely important role in current industrial design. Industrial designers appear to approach not only the meaning of products but also their functionality in terms of signs and, conversely, to regard a product's ability to bear signs as a form of functionality. The forms of products "denote" their primary function (by means of indicative functions) and "connote" their secondary functions (by means of symbolic functions). To be sure, the eventual interpretation of the sign borne by an object cannot be fully predicted. Seen from a semiotic perspective, it depends on the cultural "codes" from which embodied signs are "read." The interpretation of the primary function of products is easier to predict—though not definitively, as is clear from Ihde's concept of multistability—than the secondary function, which tends to change over time. Having pop-design furniture in the living room can be an idiosyncratic act of personal expression in the 1990s, but in the 1960s and 1970s might have meant simply that one was a "follower."

Mediation and Materiality

How is this approach, which speaks of products in terms of functions and signs, related to the postphenomenological perspective, which understands artifacts in terms of mediation? Let me say at the outset that the concept of secondary function refers to aspects of products that are closely related to what is addressed by Latour's concept of scripts. The secondary or symbolic functions of products refer, according to Muller, not only to lifestyles but also to "a certain image of the way in which, the environment in which, and the ideas through which the primary function has to be fulfilled" (302). G. Baird, for instance, in his study "The Dining Position" (Baird 1976), investigates the secondary function of tables and shows that a round table is associated with a different style of communal eating than a rectangular one. A round table has no "head" and provides everyone sitting around it with the same status, while a rectangular table is associated with a hierarchical order (Muller 2001, 303).

Nevertheless, an important difference exists between the semiotic and the postphenomenological approaches to artifacts. The semiotic perspective, that is, does not address how a table *orders* the relations between those sitting around it, but rather speaks about how it *refers* to the culture in which such relations exist. The postphenomenological perspective, however, addresses how the table *shapes* this culture, for it sees the table as actively constituting the relations between those who sit around it, whether as equally related or as related to a "head." A semiotician is not primarily interested in this mediation, but in the reference of the table to its owner's attitude toward this hierarchy. Its secondary or sociocultural function consists in this reference, not in its mediating role. Muller is implicitly aware of this difference between reference and mediation: "Cutlery, crockery, table setting, and so on, do not connote by their form only a certain sociocultural meaning of eating together. In actual interacting with them, they also condition us to the kind of behavior that is in keeping, including the way we make conversation with each other!" (303).

The conclusion seems justified, then, that the postphenomenological perspective brings to light aspects of products that have received very little attention until now by industrial design. In order to incorporate mediation into the process of conceptualizing artifacts, it cannot be treated as a new type of secondary or sociocultural function. In contrast to the sign-character of products, which can indeed be accurately described in terms of indicative or symbolic functions, the mediation of artifacts cannot be conceived in terms of functionality. Signs can be treated as means for ends: an automobile is a means not only for transportation but also for expressing one's status; a coffee pot is a means not only for producing coffee but also for exhibiting one's taste. Mediation, however, is not a product's *function,* but rather a *byproduct* of its functionality.

In fulfilling their functions, artifacts do more than function—they shape a relation between human beings and their world. The way in which a table organizes the relations between guests can only be described as a means for an end when the host deliberately chooses an "egalitarian" or "authoritarian" table in expressing a preference for a particular meal culture. But in describing how a table mediates meal culture, the issue does not concern the *function* of a table but a phenomenon that arises on *the basis of* its functionality. When the table is used (that is, when it fulfills its primary function by making it possible to lay out table settings so that people can sit in proximity), it is absorbed and incorporated into the practice of eating that it makes possible without this being consciously experienced—and from that position it mediates the relations between the people around it.

Mediation therefore does not take place in the domain of the secondary functions or sociocultural utility of products, but rather in the domain of their primary functions or material utility. Mediation, in the phenomenological sense, concerns the ways in which products function as material objects, not the ways they serve as signs. That is not to say, moreover, that products cannot also mediate thanks to their character as potential signs. As we saw, in his discussion of the speed bump and the bulky object attached to the hotel key ring, Latour notes that people's driving behavior can be mediated not only by speed bumps but also by traffic signs, while their tendency to walk off with hotel keys can also be addressed with placards. These examples clearly illustrate the difference between mediation by signs and by materiality, for the speed bump and the bulky key-ring mediate human action materially, while the traffic signs and placards do so as signs. The material form of mediation is extremely important from a phenomenological point of view, for phenomenology makes it possible to understand things as material objects that can mediate even without bearing immaterial signs.

When material mediation by things is localized in the domain of the material utility or the primary functions of products, products cannot then be completely reduced to their functionality—as was done, for instance, during the modernist movement in the history of design. The materiality of products reappears in the analysis when they are considered from the perspective of their functionality—for functionality always presupposes the material presence of a thing and not just its presence as a sign—but mediation does not coincide with functionality. What things "do" encompasses more than merely "referring" or "functioning." Things mediate the relation between human beings and their world not in a linguistic but in a material way. They fulfill their functions as material objects, and by this functioning they shape human actions and experiences.

Such "material mediation" does not take place on an interpretive level, but on a sensorial level. That a tree is interpreted differently when it is looked at through a microscope and through sunglasses is partly connected with the interpretive frameworks through which it is thought, but partly—and in an important way—also with the way in which the sensorial contact between human beings and tree is mediated differently in each case. The reason people slow down for a speed bump is connected with the concrete physical presence of the bump, which does not simply *stand for* "Slow down!" but physically compels it. Perceptions and actions always have an aspect of sensorial contact with reality, which is precisely the point of application for mediation by material artifacts.

While the materiality of products and the mediating roles they play have remained underexposed until recently, many historical surveys of industrial design begin with some remark about the huge influence that design has on social life. Thus Adrian Forty, in *Objects of Desire,* writes of "design's influence on how we think," and adds that "it can cast ideas about who we are and how we should behave into permanent and tangible forms" (Forty 1992, 6). John Heskett claims that design "radically alters the quality of life that we lead or want to lead" (Heskett 1980, 7). The postphenomenological analysis of technological mediation developed in this work makes it possible to investigate this influence more closely. This, in turn, may lead to a broader design approach, which anticipates the mediating role of things.

TOWARD A MATERIAL AESTHETICS

Since the beginning of mass production, the aesthetics of industrial production have undergone considerable evolutionary changes, from the organic references to nature of the *Jugendstil,* via the machine aesthetics of modernism, to the throw-away aesthetics of pop design and the lifestyle iconography of postmodernism. One aspect that has moved more and more into the foreground over the years, however, is the emphasis on the visual properties of objects; the aesthetics of design has grown increasingly concerned with the visual appearances of things. Under pressure of the increasing demands of marketing requirements, many companies have focused less on improving the quality of the object itself and more on its external appearance. This visualism has culminated in the post-modern obsession with signs and meanings. When products are turned into objects whose primary purpose is to refer to lifestyles, the emphasis is fully on visual qualities.

This approach understands aesthetics too narrowly in terms of "the sensory need for visual appeal and differentiation," in the words of designer Gianfranco Zaccai (Zaccai 1995, 8). Borgmann, too, criticizes the overemphasis on visualism in design: "The experiential qualities of paradigmatic technological devices such as a microwave oven, a stereo set, or a refrigerator are . . . primarily visual. The tactile and motile properties are so subordinated to ease of operation that they are nearly effaced" (Borgmann 1995, 19). Borgmann relates visualism in design here not so much to the sign-character or referential nature of products as to their functionality; they must function so well that their physical presence goes unnoticed as much as possible—they may be seen but are not to be felt. The two poles of design—functionality and meaning—thus appear to support the current emphasis on the appearance of products. This narrow conception of aesthetics, however, can be greatly broadened on the basis of a postphenomenological approach of industrial design.[3]

Aesthetics and the Senses

In saying that a certain one-sidedness is present in the aesthetics of design, I do not mean to repeat the old cliché that mass-produced objects are superficial and inferior, nor do I mean to insinuate that "genuine" aesthetics should be reserved for art and that studying the design of useful artifacts is an unworthy diversion. On the contrary, I mean to make the case for a redefinition of aesthetics in which not artworks but useful objects take center stage.

Heidegger assigned to useful objects a place midway between pure things and artworks. A useful artifact is not a pure thing, for it has to be made just as does the artwork. But neither is it an artwork, for it must be *used* in order to be present as useful. In its use it withdraws from our attention—again, in contrast to the artwork, which imposes itself on us and whose presence we linger over—though it should be noted that, as demonstrated in the previous chapters, a useful object

3. See also Verbeek (1999).

never completely withdraws, as Heidegger thought, but can evoke a certain measure of engagement with itself. In precisely such use—and also in the possibility of engaging with products during that use—lies the potential for extending the "aesthetics of things" beyond the realm of the visual. Aesthetics comes from the ancient Greek word *aesthèsis,* which means "sensory perception." Aesthetics should therefore be located in the sensory relation of human beings to the world, a relation that is not solely visual but that involves other senses as well. In certain product use, for instance, the sense of touch is at least as important as sight. After some practice, many products can be used without looking, though they could not be used at all without touching.

The aesthetics of mass products is too narrowly conceived if it is focused only on visual qualities. This would be to misunderstand how it is used, for products are then treated in effect as artworks, which are usually looked at rather than touched. The word "design," too, invites associations with art rather than with useful things (cf. Muller 2001, 145). I shall devote the rest of this chapter to a much broader conception of this aesthetics. When conceived as pertaining to the realm of the sensual and not only to the realm of the beautiful, the aesthetics of objects of use is potentially richer than that of many artworks. The sensory relations that are possible in the case of useful objects reach beyond the visual, for such things are meant to be used rather than just looked at. The aesthetics of products concerns the practical dealings with them and involves their bodily presence, rather than just what they look like or signify, or how they are to be interpreted or read.

Linking aesthetics to practical dealings with things makes it possible to draw a direct connection with the analysis of technological mediation developed in the previous chapters, which made clear that mediation occurs on the basis of practical dealings with things. When things are used, people take up a relation to the world that these things, thanks to their "handiness," coshape. In this coshaping, not only does the human interaction with products have a sensory character, so does the human-world relation that is mediated by the products. Human experience and existence can only acquire a specific shape on the basis of sensory perception and sensory dealings with the world.

Industrial design is occupied with the aesthetics of products. By extending the domain of aesthetics to include the sensorial in the broadest sense, therefore, it becomes possible to give the notion of mediation an explicit place in the industrial design process. This broadens the one-sided approach to products that sees them merely as fulfillers of functions that simultaneously refer to lifestyles. Within a sensorial conception of aesthetics, the materiality of things and of the practical dealings with them—rather than only looking at them as if they were artworks—become relevant again. The meaning of aesthetics in design then comes to include not just style and beauty, but also the relations between people and products, and the ways in which products coshape the relation between humans and world.

Aesthetics and Ethics

To give such a broad scope to aesthetics, in which the mediating role that artifacts can possibly play in their context is explicitly anticipated, also entails that an ethical dimension enters into product design. Ethics is concerned with questions of how one should act (ethics of behavior) as well as with the more classical question of "how to live" (virtue or life ethics). And now that it has become clear that artifacts play a mediating role in human action, they appear to provide a "material answer" to this question.[4] Designers engage in "ethics by other means"; that is, their products codetermine the outcome of moral considerations, which in turn determine human action and their definition of "the good life." I shall first discuss a number of ways to approach the moral dimension of artifacts and then engage in a critique of them. For the coupling of materiality and morality usually raises the question whether the influence of things on people should be described in terms of conditioning and steering rather than in terms of genuine morality.

Bruno Latour was one of the first to speak explicitly of the mediating role of artifacts in connection with ethics. In his article "Where Are the Missing Masses?" he describes with characteristic irony how the concern for the supposed moral decay of modern life—the "missing masses" of morality that excite contemporary politicians as much as the question of "missing mass" excites contemporary astronomers—can be solved when the missing masses are sought after not solely amongst humans but in things as well (Latour 1992). The emblematic example that he develops here is the car that is programmed not to start, or to emit an annoying series of beeps, if the driver has not fastened the seat belt. Such an automobile is full of morality, for to its machinery has been delegated the task of enforcing the moral determination of whether or not the driver should wear a seat belt. Concerning that moral issue, the spirit is willing but the flesh is weak for many people. In Latour's example matter supplies the necessary force, whether in the form of shutting down the automobile outright or emitting an irritating noise so that the choice becomes easy.

4. See De Vries (1999), 15–30.

While Latour implicitly localizes the moral dimensions of things in the domain of the ethics of *behavior*—for him, things help to shape the answer to the question of how one should act in a given situation—Gerard De Vries connects the mediating role of things with classical *virtue* ethics. In this tradition, from Socrates onward, ethics turns on the answer to the question, "How to live?" This question is not formulated in terms of morally responsible subjects who make choices, but in terms of the kind of life to be led. According to De Vries, in a technological culture the answer to this question is partly provided by the many devices amidst which human beings live. "It is not just ethicists or theologians who answer this question. The landscape as well as the city are both highly structured, and our existence is furnished with many different kinds of devices and technological systems. *These* are what instruct people in contemporary societies 'how to live'" (De Vries 1999, 15–16). Ethics consists for De Vries of "a combination of empirical and philosophical investigation" (29); a philosophical analysis of situations in which (material and nonmaterial) answers are given to the question of "how to live" on the basis of a thorough study of empirical reality.

The Dutch philosopher Hans Achterhuis has proposed to make this notion of the "morality of artifacts" fruitful for the design of technology, suggesting that designers should explicitly anticipate it. In Achterhuis's opinion, we should start moralizing our artifacts, not just each other. In order to prevent people from continually having to reflect on their actions, which would amount to a paralyzing decisional burden, some decisions should be delegated to devices (Achterhuis 1995). The speed control device in an automobile is a good example. To make our highways safer, we can strive to change the mentality of automobile drivers, but one wonders how much change we can really expect given the availability of highways that make it easy to drive much too fast and cars that are able to exceed the speed limit by far. What is the best way to begin if we want to lower the number of traffic fatalities? "In 1991 300 automobiles collided in dense fog on the A16 highway in The Netherlands, with ten fatalities. They were driving much too fast. If automobiles were forced automatically to slow down to a safe speed when such fog occurred, the fatalities certainly would not have happened." Achterhuis ironically adds: "But in this case, the freedom and responsibility of the drivers were preserved" (Achterhuis 1998, 379).

Many objections have been raised to the call for a "material ethics." Achterhuis, for instance, has been accused of wanting to establish a technocracy, and his proposal to delegate the enforcement of human choices to devices has been accused of leading to totalitarianism. What will remain of human freedom and dignity if we voluntarily submit ourselves to rules imposed by devices? (cf. 28–31). But such objections ignore Achterhuis's opposition to conditioning human beings behind their backs, so to speak, and his insistence that the "moralizing" of our artifacts should result from public discussion. I would add that it is precisely those who refuse to engage in public discussion about the moral valence of artifacts who subscribe to a totalitarian approach. Artifacts always play a mediating role in our lives, and shunning public discussion of their influence amounts to leaving the answer to the question "How should we live?" up to the engineers—which would indeed be the technocratic solution. Apparently the solution to the problem of the "missing masses of morality" is still too unfamiliar to handle responsibly. Moreover, the mediating role of artifacts does not need to consist in conditioning or force; more subtle and less controversial forms are possible, as I shall argue shortly.

A second objection to assigning morality to things concerns the fact that artifacts are not able to make deliberate decisions about their influences on human action. Things lack intentions, runs this objection, and thus cannot be held accountable for their behaviors. This in turn makes it unreasonable to characterize their influence on human action in terms of morality; it is just plain influence. Tsjalling Swierstra is a representative of this critique, which he summarizes in his claim that artifacts cannot fully partake of the "moral community." The definition of this community has been considerably expanded since ancient times, Swierstra argues. "Women, slaves, and barbarians were largely or completely excluded from moral rights," though "in the course of time all of these groups have been admitted" (Swierstra 1999, 317). But the trend now toward admitting things as members of a moral community goes too far for him—at least in the most conventional form of ethics, the ethics of behavior, which seeks to answer the question of how people should act in certain situations (321–23).

This ethics of behavior comes in two varieties, consequentialist and deontological ethics. Consequentialist ethics evaluates behaviors exclusively on their (likely) consequences. An action can be characterized as moral when its (likely) positive consequences outweigh its negative ones. From this perspective, where it is only the result that matters, things can indeed play a role in moral practice, for they can incite people to morally right behavior. But things do so because humans tell them to do so. They themselves cannot evaluate the (likely) positive and negative consequences of their influence on human actions. From a consequentialist perspective, therefore, things are only moral instruments rather than independent moral agents.

In contrast to consequentialist ethics, deontological ethics focuses not on the consequences of an action but on the moral value of the act itself and the intentions behind it, regardless of the consequences. From a Kantian perspective, for instance, the morality of an action depends on whether or not the agent has sought to act in accord with rational norms. Artifacts, of course, cannot themselves make such evaluations. Moreover, if artifacts incite humans to act in a way that is desirable from a deontological perspective, then their actions at that moment do not arise as a rational duty but are simply steered behavior.

Thus from both a consequentialist and a deontological perspective, according to Swierstra, things at most can be held causally responsible for an action, never morally responsible. Nobody would ever summon an automobile to appear in court for its role in an accident. "Coercive artifacts are themselves not moral actors, nor can they produce moral behavior in humans. Thus . . . no grounds exist . . . for admitting artifacts as members of the moral community" (Swierstra 1999, 322).

Swierstra's reasoning, however, involves a remarkable conceptual leap. He defines moral community in terms of what can claim moral rights; the members of the moral community are those who can *lay claim to* moral treatment (317). But when he rejects things as being potential members of such a community, he does so by stressing their inability to actively *shape* behavior or *embody* morality. In the first case he treats them as receivers, in the second case as carriers of morality. His position that artifacts cannot participate in the moral community is only valid if carrying moral responsibility is required to make a claim to moral treatment—and that is a problematic assumption. Were that so, children would have no moral rights, and environmental ethics would be impossible.

Swierstra implicitly tries to avoid this problem by shifting his attention to care ethics, which recognizes asymmetric relations such as those between adults and children, and which offers him a way to grant some moral status to things without then having to recognize them as full-fledged moral agents. The development of a "care ethics of things," according to Swierstra, would allow us to accept things as involved participants of a moral community, though "in a limited sense," and "not because they moralize us, nor because they are responsible agents, but because they (can) become objects of care" (325).

Swierstra thus sticks to his definition of moral community in terms of the ability to make a claim to moral treatment. But to divorce the carrying of morality and the claiming of moral treatment, as happens in care ethics, can also be taken in another, quite different direction. Moral community can also be defined in terms of the ability to *shape* morality rather than in terms of the ability to *claim* moral treatment. It can be seen, that is, as the community in which answers are given to Socrates' virtue-ethical question, "How to live?" or to the ethics of behavior question, "How should I act in this particular situation?" And this would make things full participants of the moral community. Even more strongly, defining things this way would make it clear that ethical reflection cannot take place adequately if it ignores the moral valence of things.

I am fully in accord with Swierstra's reluctance to accept a thoroughgoing symmetry with respect to the moral status of human beings and things, as I indicated in Chapter 5 in arguing against Latour's symmetry and for the phenomenological differentiation between humans who act and a world of things in which action takes place. Things do not have intentions and cannot be held responsible for what they do. But that does not alter the fact that they do act. They play a mediating role—one with an ethical dimension in that moral considerations are transformed, shaped, or even taken over. This mediation of moral considerations can be explicitly delegated to things, as Latour and Achterhuis illustrate with examples such as speed bumps and speed controls. But they can also arise implicitly, as is the case, for instance, with prenatal screenings that make possible the discovery of diseases in fetuses and can implicitly invite abortion (Popkema et al. 1997). In all these instances the things themselves do not do the moral evaluating, but they do contribute to it—which makes it worth the trouble to anticipate this contribution. If that anticipation does not happen, things would have free play in answering our moral questions, since nobody would try to adapt their built-in morality then. It is precisely the failure to recognize that things have a moral valence that gives rise to technocracy.

But serious ethical attention to things is necessary not only because things mediate the moral considerations of people. A second reason is that the mediating role of things can be judged in moral terms itself, whether this role has been explicitly delegated to things or not. The interesting issue is not whether the city planner Robert Moses himself was elitist or racist, but rather whether the impact of his overpasses on the highways to Long Island beaches was elitist or racist by the way they deliberately prevented buses from passing and allowed only automobiles, which were typically owned by members of the white middle class (see Chapter 3). The fact that the overpasses themselves cannot be blamed for this elitist or racist practice does not mean that their impact on practical life cannot be evaluated in moral terms. When designers do not pay attention to this mediating role of things, ethical reflection in the design process remains incomplete. Design ethics requires that artifacts be treated as members of the moral community, conceived as the community in which morality assumes a shape. Things carry morality because they shape the way in which people experience their world and organize their existence, regardless of whether this is done consciously and intentionally or not. The very fact that they *do* this shaping charges designers with the responsibility to make sure that things do this in a desirable way.

These considerations show, therefore, that design has two types of moral dimensions. First, designed products play a mediating role in the moral considerations of people, and second, the design process can involve moral choices with reference to this mediating role. But anticipating mediation brings up a complicating factor. Earlier in this book I wrote about the "multistability" of artifacts—the fact that they lack an essence, and are what they are only thanks to the contexts in which they play a role. For this reason, the mediating roles of artifacts are not properties of the artifacts themselves, but arise in the relations which people have with artifacts. In the Netherlands, for instance, the automobile only came to be thought of as a means for long distance transportation after the great rail strike of 1903. It was initially used chiefly by hobbyists and for medical purposes: "to inhale fresh, pure, dust-free and—thanks to the high speeds—rarefied air, and the climate changes that accompany car driving were valued as exceptionally beneficial to the lungs" (Baudet 1986, 78). When the Post Office became incapacitated by the strike, however, it called upon automobile enthusiasts to help deliver the mail—and suddenly the automobile came to be perceived as an alternative to the train for long distances (74–84). And only thanks to this way of interpreting the automobile did it come to have the role it now plays, as something that shapes our choice of dwelling and workplace, the way we take our vacations, the frequency with which we maintain our social contacts—and even, it is said, the number of children we want to have, as can be derived from the available space in the back seat (Swierstra 1999, 329). This multistability of things makes it difficult to anticipate the eventual character of the mediation, and thus to explicitly anticipate it in the design process. But this anticipation is not impossible. The problem presents itself particularly in the case of entirely new product categories. Wherever conventions are already in place concerning particular objects, some stability has arisen in the multistability. Within design theory, extensive attention has been paid to such "stability," with constant research into the habitual use of particular products and into the degree to which particular product forms are in fact used for an intended end. Such research is already a significant part of design processes (Muller 2001, 287–332). Thus the existence of multistability—a product's "interpretive flexibility," as Bijker calls it (Bijker 1995, 20)—need not hamper designers in explicitly trying to anticipate the mediating role of products in their use context.

DURABLE DESIGNS

To demonstrate the practical relevance of the material aesthetics discussed above, I shall elaborate how it can contribute to the discussion of ecologically friendly industrial design.[5] I shall make clear that this discussion can be fruitfully supplemented by a perspective in which the relation between humans and products takes center stage, instead of focusing separately on humans (who should act more friendly toward the environment) and things (which should be "cleaner").

For some years now in the Netherlands an industrial design organization called Eternally Yours has been active, whose goal is to generate ideas for addressing ecological product development. Eternally Yours seeks to develop durable products, however, in an unorthodox way. The usual approach to "ecodesign" is strongly instrumental and technological in nature. It begins with lifecycle research and analyses, so that the ecological impact of a product can be evaluated in all stages of its life span—production, distribution, use, and discarding—and takes into account maintenance, repair, and reuse as well (Te Riele and Zweers 1994, 23). But Eternally Yours sees this approach as too limited, for it neglects the important fact that most products are thrown away long before they become obsolete. As long as this happens, it does not matter how "clean" a product is. It is at least as important to try to get people not to throw away their things so quickly.

In seeking to achieve that goal, Eternally Yours tries to give products properties that allow humans to develop attachments to them. Such an approach requires a completely different vision of technology than is available for conventional environmentally friendly design, for it is oriented toward the relation between technological products and their users rather than toward emissions, available resources, and energy consumption. The postphenomenological perspective and material aesthetics that I have been developing are highly relevant to such an approach to ecodesign, for they allow designers to approach human habits concerning product disposal as something wherein the products themselves play an active—and therefore changeable—role. Artifacts inevitably mediate the relations between humans and their world, and therefore also the relations that humans have to the artifacts themselves. By consciously anticipating this mediation, in which sensorial aspects will appear to play an important role, it may be possible for designers to inscribe in products an "antidisposal ethics."

To be sure, Eternally Yours does not claim to provide a full alternative to ecodesign, nor is it saying that its products take away the need to radically change

5. An earlier version of part of what follows appeared in Verbeek and Kockelkoren (1998).

our environmentally demanding lifestyle. Environmental problems cannot be entirely solved by persuading people to hold on to their stuff longer. The point is rather to develop a new perspective on these problems, which focuses not just on people who should be moralized and change their lifestyles, nor just on products, which should get a cleaner lifecycle. In this new perspective, the dichotomy between people and things is overcome precisely by taking into consideration their mutual relations.

Eternally Yours

Existing approaches to environmentally sound product design, as just noted, seek to eliminate bottlenecks created by products in the various stages of their lifecycles. The way in which the environmental burdens of products are usually addressed tends to focus almost exclusively on pollution and the use of natural resources. Thus in the Netherlands, the government-sponsored program for ecodesign has adopted the so-called MET approach, in which MET stands for "Material cycle," "Energy consumption," and "Toxicity." The tripartite MET agenda is as follows: "To optimize aspects of the material cycle throughout the entire lifecycle (maximize time of material use, encourage high-quality reuse, cut down on the amounts of material used); reduce energy consumption throughout the entire lifecycle (both the time and amount used); eliminate or reduce toxic (and nontoxic) emissions in the entire lifecycle" (Te Riele and Zweers 1994, 29).

The results of the MET approach do show that it can indeed make great inroads into reducing the environmental impacts that products have, but it does not address the issues indicated by Eternally Yours. The MET approach does take into account waste disposal, but it does not seek to prevent or postpone the throwing away of products, but rather to minimize the amount of environmental damage created by this disposal. To be sure, within the MET approach a fourth directive is added, as Te Riele and Zweers indicate: "Increase product quality (efficiency, effectiveness, and duration)." But the attention for product longevity that is proposed here is not firmly rooted in the MET approach. This is illustrated by the fact that no Q for "quality" or L for "longevity" earned a place in the acronym to indicate the approach. Besides, in the so-called MET-matrix, which plays a central role in the MET approach (indicating the material use, energy use and toxicity in the various stages of a product's lifecycle), product quality does not play a role of any importance.

As the work of Eternally Yours shows, this Dutch ecodesign project misses important opportunities. Eternally Yours is concerned about the growing stream of superficial products that replace each other ever more frequently, and that are designed to be quickly discarded even while they still function quite well. It is interested not only in explicitly disposable products such as plastic cups and tableware, but also in the growing number of products that, as studies have shown, wind up in the landfill even though they still operate well or could easily be repaired (see figure 4).[6] Some products do not even reach the market anymore; recycling companies regularly destroy brand new hard disks, which are outdated already before they can be sold. Sustainable designs should be durable, according to Eternally Yours: designers should first of all lengthen the life span of products. This is not an obvious strategy. Eternally Yours identifies four possible ways to address the environmental problems connected with mass consumption (*Eternally Yours* 1997, 3). The first way is ecodesign by means of lifecycle analyses. A second is to bring about a shift from products to services. Expanded repair and upgrade services, for instance, as well as agencies such as lending libraries and car sharing, can increase product use in a way that would minimize the number that would have to be produced. A third option is recycling. Reusing parts of discarded products can reduce the consumption of energy and raw materials. But these three ways do not address head-on the problem of product disposal and overconsumption. The fourth way is therefore to try to extend the service life of products.

Eternally Yours discerns three aspects of a product's life span: technical, economic, and psychological. Products are thrown away for these reasons: because they simply break down or cannot easily be repaired; because of economic reasons, such as the appearance of newer models on the market; or because their owners have changed their preferences or self-images. For Eternally Yours, this last reason is of particular interest and is the most important, for the majority of products appear to be thrown away because their psychological lifetime has been exhausted. The most relevant issue for durable product development is therefore to devise ways to lengthen this psychological lifetime. How can stronger bonds, that is, be fostered between people and the artifacts around them? Seeking answers to this question calls not only for an instrumental-technological solution to environmental issues but also for the generation of what Eternally Yours calls "cultural durability."

Eternally Yours draws inspiration for the content it gives to cultural durability from the Italian designer Ezio Manzini. In a quote that appeared as the letterhead of Eternally Yours for some time, Manzini said, "It is time for a new generation

6. This figure is taken from Eternally Yours (1997, 4) and was drawn up by Eternally Yours on the basis of results of research by Blonk (Bureau B&G, Rotterdam) for the Dutch Ministry of Environment. The survey involved 20,000 respondents and took place between 22 September and 31 October 1992.

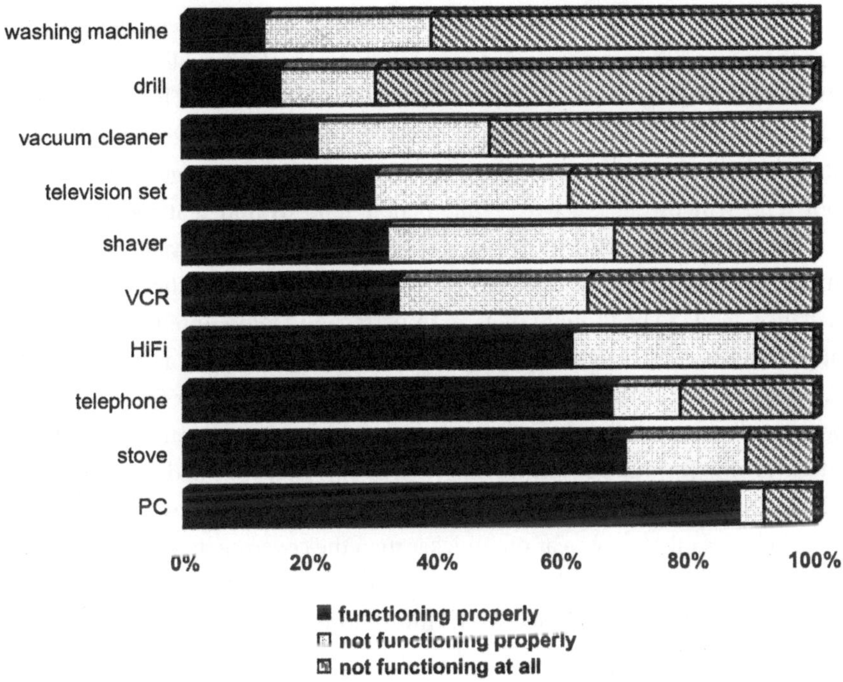

Fig. 4 Discarded products and their condition

of products, which can age slowly and in a dignified way, can become our partners in life and can support our memories." Manzini is concerned about the state of our material environment. "Today we live in a world of objects designed for rapid consumption, objects requiring a minimum of effort and attention to use them, but also objects that leave no lasting impression on our memories—a throw-away world that requires no effort but, at the same time, produces no real quality." Designers, he said, should be guided by the image of objects as plants in a garden; they should see themselves as planting "a garden of objects." Just as do trees and plants, objects have "lives of their own"; they "perform services and require care." Manzini makes a plea for "caring for objects," with the word "caring" bringing with it a new "ecological sensibility," namely, that "caring for objects can be a way of caring for that larger object that is our planet" (Manzini 1995, 222, 239).[7]

7. For a more philosophical approach to "caring for objects" from the perspective of the designer, see Fry (1994, 122–33).

How, then, can the psychological lifetimes of products be lengthened? How can the bonds between people and products be fostered? Eternally Yours has formulated three projects in seeking answers to these questions. The first, referred to by the rubric "Shape 'n Surface," seeks to increase products' lifetimes through their forms and materials, such as by seeking materials whose aging process does not render them unattractive. While people tend to find that the surfaces of leather or wood age attractively—and the same is true of a number of synthetic materials—the first scratch or tarnish on the surface of polished chrome, for instance, tends to be thought of as an imperfection (Van Hinte 1997, 126). This approach has led to some surprising results. Thus Sigrid Smits, a graduate of the Academy of Industrial Design in Eindhoven, has explored for Eternally Yours ways of allowing furniture to age attractively. She has incorporated patterns, for instance, into furniture coverings that emerge only over time and with wear (Van Hinte 1996b, 21–22). She has also put seams on the surface of the coverings rather than below, so that they wear out quicker than the covering itself and in wearing out develop a progressively more expressive character (Van Hinte 1997, 130). Such design ideas allow a piece of furniture to look renewed over time rather than worn out.

A second project for increasing product lifetime is generated by the creation of possibilities for improving the support and repair of products. This goes under the rubric of "Sales 'n Services," and concerns the organization and deployment of services such as cleaning and repair as well as other services and activities that could enhance product longevity. Businesses could shift their emphasis from simply producing and selling products to establishing and maintaining a relationship with clients. The stronger this relationship, the more extended the survival time of the product around which this relationship takes place is likely to be.

Under the rubric "Signs 'n Scripts," finally, Eternally Yours investigates how the sign-character of products and their "scripts" can be used to extend their service life. Products can feature in stories, serve as icons for lifestyles, and contain implicit instructions for use. All these aspects can play a role in their service lives: stories can give products more character, and their significative value can allow us to experience them as relevant to ourselves. (I shall discuss below the way in which scripts can influence a product's service life). To illustrate this dimension of product lifetime Eternally Yours cites an ad campaign used by Nikon a few years ago. One spread featured a camera that had lain underwater on the sea floor for ten years but still worked. The ad created a context for the product advancing its psychological lifetime, for in the camera's story it did not play the role of the newest model with the latest gadgets—which would have to be replaced once a still newer model hit the market—but of a solid camera of which its owner could be proud.

Attachment Between People and Things

The aim of Eternally Yours to find ways to design products so that people acquire stronger attachments to them invites a closer analysis of the relation between people and products. When do we become attached to things? Mihalyi Csikszentmihalyi and Eugene Rochberg-Halton have conducted empirical research into the meanings that products have for their owners. This research, described in their book *The Meaning of Things: Domestic Symbols and the Self,* brings to light interesting aspects of the relation between people and objects. Those surveyed identified their most special objects as: furniture (mentioned by 36% of the respondents), visual art (26%), photographs (23%), books (22%), stereo equipment (22%), musical instruments (22%), and television sets (21%) (Csikszentmihalyi and Rochberg-Halton 1995, 58).[8] The most important reasons the respondents advanced as to why they found these objects special seemed to be: the memories that clung to them (with respect to photographs this was mentioned by 26.7% of respondents; to visual art, 15.6%; furniture, 15.4%); the experiences connected to the use of these objects (television, 31.5%; stereo, 28.1%; musical instruments, 27.1%); and reference to immediate family (photographs, 26.0%; musical instruments, 17.1%; visual art, 15.6%) (88).

These data allowed the authors to develop a conceptual framework in order to understand how objects acquire meaning for people. The authors proposed that meaning was generated by the active interaction between people and things, which they interpret in terms of a transaction process. Transactions are conceived as "psychic activities (or communicative sign processes) and not simply physical behaviors per se, although they involve physical behaviors" (175). Transactions between people and things consist of three elements: experiences of the *aesthetic quality* of the object, the *attention* devoted to the object, and the *goals and outcomes* of the transactions. Objects can acquire their meanings for people first of all from their intrinsic aesthetic qualities. That is true, naturally, for artworks, but also for useful objects in general to the extent that people find their exteriors pleasing or that they express a particular lifestyle with which people want to be associated. Next, objects can acquire meaning by the ways people deal with them. The authors speak about such "attention to things" in terms of what they call

8. I identify only those objects mentioned by more than 20% of the respondents.

"flow." Hi-fi equipment, televisions, and musical instruments, for example, allow people to participate in what they make possible. Someone who sits on the stool before a drum set is "in" the music; and even someone who watches a film is "in" the story. Finally, things can contribute to or express what people hold to be their highest values or goals in life. People display photographs to express how important family is in their lives, not so much because of the aesthetic quality of the photographs or the "flow" that they make possible.

The studies by Csikszentmihalyi and Rochberg-Halton illuminate the reasons why people attach themselves to objects, but do not yet provide a completely adequate conceptual point of departure for the kind of work that Eternally Yours wants to do. Asking people which objects they find most special invites answers that concern the objects to which they have the most emotional attachment. The most common answers, for instance, concerned objects to which memories were associated and that referred to family members: photographs and heirlooms such as furniture, musical instruments, and art. But such intense emotional bonds cannot be demanded from all the things and artifacts to which people are attached. Manzini himself has remarked on this: "One cannot possibly feel attached to each and every product. I don't agree that all products must be Eternally Yours" (Van Hinte 1997, 234).

A perspective on "culturally durable products" thus should not take a "what would you take with you to a deserted island" approach. Cherished possessions do not offer the most suitable model for fleshing out the kind of "cultural durability" that Eternally Yours is seeking, for it cannot be expected that every product will become irreplaceable thanks to the memories that will grow up around it, or to its family references. Moreover, it is difficult for designers to incorporate such aspects into products because they arise in the course of the interactions that people have with them.

Culturally sustainable product development should not aim at "devotion" to products, but at attachment. Products to which people develop an attachment are not generally as emotionally charged and irreplaceably present as heirlooms, but neither are they as anonymous as a throw-away item. Not seeking a model in irreplaceable personal possessions creates the space needed to direct attention to more everyday products, of which so many wind up most often in landfills. What distinguishes these goods from our most loved possessions is that they are *used* rather than *cherished*. Most photographs sit tucked away in albums, and artworks hang on walls—but coffee makers, heaters, benches, and computers are used daily. The challenge for designers lies in evoking an attachment with the user on the basis of this use.

Of the reasons cited by Csikszentmihalyi and Rochberg-Halton for the attachments that people have to their cherished things, only the "experiences" that belong to such objects have to do with their use. Such experiences offer an important point of connection to the project of Eternally Yours, as I shall soon show. They concern practices that spring up around the materiality of products and that have a sensorial dimension, which means that they take place in the domain of the "expanded aesthetics" sketched out above. But when the dimension of experience is more closely analyzed in terms of the transactional model which they propose, the sensorial dimension does not fully emerge. Csikszentmihalyi and Rochberg-Halton conceive transactions between people and things as "psychic activities" or "communicative sign processes." And of the three aspects of transaction (aesthetic quality, attention, goals), only "attention" refers to objects in use—but this in turn is understood in terms of "flow," an especially intense form of involvement that does not apply well to most everyday, useful objects.

In order to prevent people from throwing away objects when a newer model appears on the market, when the prevailing fashions shift, or when they need repair or maintenance, connections must be forged with other aspects of product use. These consist in the mediating ability that artifacts possess. For realizing the aims of Eternally Yours, this mediation must take place in a way that stimulates an attachment between people and the artifacts themselves as material objects. The bond that arises between people and products will have to concern the concrete object that is present in the here and now, and not only the meanings or symbols it carries or the functions it fulfills. If someone's attachment to an object is only based on the way it expresses his or her lifestyle, then the object is vulnerable to being replaced by any other one with the same sign characteristics. The same holds true if the attraction is based only on the functionality of products, their roles in stories, or the fact that they may serve as an introduction between a business and its clients. To enact true "cultural durability," human involvement with objects cannot have merely a nonmaterial orientation, but must be oriented to the material object itself. But how can artifacts mediate the relations that people have with them in this way?

Transparent Artifacts

An initial phenomenological point of entry into the ways in which artifacts can mediate the bonds people have with them through their use is provided by Heidegger's distinction between readiness-to-hand and presence-at-hand. That distinction can be used to characterize the way products are present in a use context. When artifacts are used, they are ready-to-hand, as Heidegger says, for they make a practice possible without themselves becoming objects of experience or action. Handy objects withdraw from the field of human intentions in order to make possible a relation between people and their world. When artifacts break down, they become present-at-hand. The practice that developed around them ceases, and they now become objects of experience and of action.

A first suggestion that arises thanks to this distinction might initially seem trivial. Objects that are used have to be ready-to-hand, which implies that in order to have a durable relation with products it is necessary that they allow a return to readiness-to-hand after becoming present-at-hand. The problem with many products is that such a return to readiness-to-hand is not possible, because they are literally and figuratively not transparent enough for that. For example, many plugs and power adaptors for electronic products are tightly sealed instead of being fastened in place in a way that would allow them to be opened for repair if they fail to work. Once the ready-to-hand relation with such products is broken, nothing can be done with them. In order to make a return to readiness-to-hand possible, products have to be *transparent,* in the sense of being devoid of obstacles that stand in the way of our being able to restore their functioning and return them to the place they held in our dealings with them. Their functioning should be understandable and accessible. A tightly sealed covering forms an impenetrable obstacle, making the object opaque. Transparency is necessary not only to make products reparable and thus to extend their physical lifetime, but also for extending what Eternally Yours calls the "psychological lifetime" of products, for it facilitates recovering relations with them even after a disruption therein.

An absence of transparency, of course, can arise in other ways than by sealing up the product. Even products that can be easily taken apart are often not easily repaired and do not readily promote a relation with their workings. Notes Ed Van Hinte, "If you try opening up the insides of your television set, you inevitably find, in the middle of a jumble of incomprehensible parts, a dirty sticker that orders you not to touch anything or you might get an electric shock. The interior of useful products has degenerated from a place of aesthetic and technological grandeur to a stuffy minefield that belongs to someone else" (Van Hinte 1996a, 29). Van Hinte's observation reveals that many products are designed with two separate territories: the covering or "skin," which is freely available for users to look at and touch; and the interior, which is only for trained technicians to access. "This segregation into territories is quite peculiar. The person who buys a vacuum cleaner or a walkman does not have the entire purchase at one's disposal. Yes, you can unscrew it and open it up—if you understand how to do this, that is, which might be quite a puzzle—but your attempt will be punished by loss of your guarantee, and in the case of some larger appliances you may even be threatened with death" (29).

Van Hinte describes how this territorial segregation arose in industrial production and especially in the emergence of electronics, in connection with which a steadily increasing amount of technical knowledge was required to understand the devices in question. He sees the development of the Apple Macintosh computer as bringing the separation of a product's interior and exterior to culmination. This computer was designed so that it would be as easy as possible to use, while the technology thanks to which it was possible was made completely inaccessible. "Steve Jobs and Steve Wozniack sealed their territorial conquest by working their signatures into the inside wall of the Mac housing" (32). But meanwhile their creation became impossible to upgrade—which is essential for computers, for rapidly developing software continually raises the standards for new programs and requires ever more powerful hardware—and to repair when it malfunctions.

A consequence of this inaccessibility of product insides is that it does not allow the development of an adequate relation to the products themselves as a material objects, and therefore discourages attachment. Van Hinte proposes that this be changed by encouraging what he calls "functional clarity" with respect to the individual components of product insides. This would make clearer to the owners of these products what these key components are and what they do, and how to repair or replace them. Such "functional clarity" fits extremely well with the thought that objects must be made "transparent" in order to make possible their return from objective presence to handiness. A beautiful example of an object that possesses such transparency is the "Ithaca" Color Printer, which Donald Carr designed at the Cranbrook Academy of Art in Michigan (see figure 5). This printer does not cloak its insides but openly exhibits them, and makes clear what role is played by which part. It is not a black box that simply works or does not work. Transparency makes attachments between people and products possible in two ways. First, it allows people to maintain a relation with products even when they break down. Second, and more important, it makes it possible for people to become involved with products as material entities. For when a product is transparent, it is not only functionally present but it exhibits *how* it is functioning. This is true in connection with what the product does for its user—its *commodity*, in Borgmann's terms—as well as with the way to that result—its *machinery*. For attachment with the product can arise only when the machinery of the product makes involvement possible. If it is only a question of obtaining commodities, it does not matter what object provides the commodities.

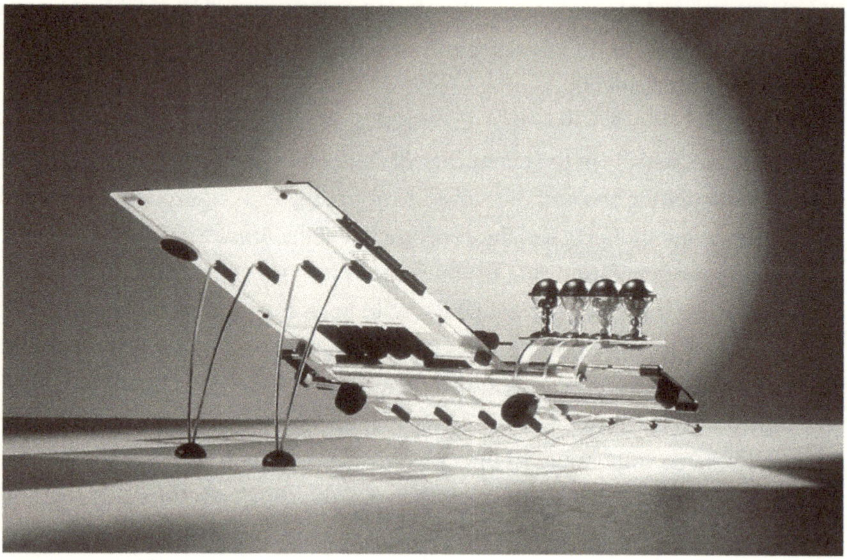

Fig. 5 "Ithaca" color printer (Design: Donald Carr)

In Borgmann's terms one may say that the design of transparent products is a way of stimulating *engagement.* By this I do not mean the focal, meaning-giving engagement of which Borgmann speaks, but rather bodily-sensorial involvement with objects. Transparent products break through the technological pattern in which the machinery of devices recedes into the background and involve people with their machinery. They do this first of all by making perspicuous *how* they function, though such involvement has primarily a conceptual dimension, for people are not actually bodily involved with the function of transparent products. Second, they break through the technological pattern by making repair and revaluation possible, and in that way they invite bodily-sensorial involvement with themselves. Transparency is thus one of the characteristics of the aesthetics of culturally durable products.

Engaging Artifacts

Products can be engaging in more ways than just by being transparent. After all, it is much more obvious to locate human involvement with artifacts in their use rather than in their repair or upgrading, as is the case with transparent products. By the term "engaging products," I mean to refer primarily to products that involve humans in their functioning—rather than in retrieving or extending their functionality, as do transparent products. Engaging products, too, can be understood via Heidegger's distinction between readiness-to-hand and presence-at-hand. In Chapter 6, it became clear that engaging products are present to human beings in ways that are neither entirely ready-to-hand nor entirely present-at-hand. While a ready-to-hand artifact completely withdraws from the relation it makes possible between humans and world, an engaging artifact remains explicitly present in that relation, but without demanding so much involvement that it becomes present-at-hand.

In the previous chapter this form of presence was illustrated with the example of a piano, which calls for active involvement in order to produce music, but which at the same time is "embodied" and ready-to-hand. But more ordinary useful objects can also be used to make this point. Many objects are designed so that they are merely functional; their machinery withdraws so that only their functionality appears in the foreground. A good example of this is the sealed electrical power adaptor mentioned above. In use, it is taken up into a human activity. When a Walkman is turned on, the adaptor is not noticed so long as the music continues. But the way this adaptor, and many other useful objects, withdraws in its functionality does not lead to a durable relation with it. Its presence is purely functional in nature. It withdraws from human involvements so that people have no relation to it as a material object but only as something that fulfills a function. It works, and that is that. If it breaks, you buy another.

To be present in a more than functional way, products could allow people to become involved in their functioning, in the way, for instance, that a piano only produces music in active interaction. To allow the users of a product to participate in the way in which it delivers its "commodities" would be to involve them in its material machinery, so that their relation to it would not be limited to the thoughtless consumption of what it makes available. In Latour's vocabulary, the engaged possibilities of products would be described in terms of the delegation by nonhumans to humans. In order to invite the involvement of users, the responsibility for certain aspects of their functioning must be handed over to people. People would become, as it were, part of the machinery, for their actions and involvements would be translated in a way that would involve them with the materiality of products, and not only with their functionality or meaning.

For the designing practice, this implies a rehabilitation of the machinery of products. In order to involve people with products as material things and not only with their meanings or the lifestyles they represent, products must be designed that are more dependent on humans rather than functioning quasi-autonomously. Products that allow human participation in their functioning, or with their repair when they break down, forge a bond between users and themselves as material things rather than simply as suppliers of commodities. The electrical power adaptor functions quasi-autonomously, for it requires no looking after when it does its work. Once it is properly plugged into an outlet, our involvement with it ceases. A piano, by contrast, requires human involvement in its functioning. When someone plays a piano, that person is involved with the piano as a material thing and not only as a supplier of sound.

A piano, however, is a rather easy example of an engaging product. Pianos, after all, are meant for active music making, and a piano not destined for this form of engagement would be unthinkable. What might the notion of engagement mean for the design of objects that are not intended for involvement at the outset? An example might be a heater, which Borgmann uses as a textbook example of a device that invites consumption of a commodity rather than engagement. At the Cranbrook Academy of Art, Sven Adolph has designed an electric space heater that breaks all conventions surrounding the design of such devices. It consists of a set of ceramic plates that are bent around each other concentrically, tapering off at the top. Each cylinder is open at the top and also has a lengthwise opening. An attachment allows the different elements to be shifted around, so that the heat can be radiated in different directions.

This product explicitly involves users in its functioning. It is not an object to be tucked away beneath the windowsill, as are most heaters—it belongs in the middle of the room. To be used, the position of the elements has to be adjusted and the design of the heater encourages people to sit around it like a campfire. The possibilities to which this heater gives rise invite involvement and not just heat consumption. It is functional, but its properties are not limited to functional ones. Not only the fact *that* it functions is important, but also *how* it functions. The space heater is ready-to-hand, but at the same time solicits our involvement without becoming burdensome or opaque. Not only does it call attention to itself; it also succeeds in comfortably warming the room.

We maintain a relation with products designed in such a manner not only when we turn them on or plug them into an outlet. This space heater therefore illustrates well the material aesthetics I elaborated above. It is not only visually attractive, but invites a broader sensorial relationship, which lets it be present as a material object, mediating the relation that people in the room have to each other and to the space heater itself. And although radiators are not objects that we often throw out, Adolph's space heater clearly illustrates the ways in which products can shape the attachments that people have to them. Products such as this one are absorbed into a practice without making impossible a relationship to it as a thing. Their presence to us is not exhausted by their functionality, for we remain involved with them as material objects.

Another example of a technology that involves humans in its functioning, though not specifically designed for that purpose, is the wind-up radio manufactured by BayGen. This radio was specially designed for use in developing countries where electricity is not always or everywhere available. It does not require batteries, but is powered by a generator that is driven by a wind-up spring mechanism. Thus it calls for regular attention to itself just like traditional spring-driven wrist watches, which also from time to time must be wound. To receive a radio broadcast much more is thus called for than setting a dial.

A second way in which products can call for involvement by virtue of their use is not so much through human involvement in their functioning, but instead through their integration in everyday practices. Products toward which people have only an indifferent use relation could be redesigned so that people can make them their own. The computer is a good example, which was designed to function in a business context and to be at home, so to speak, in an office rather than a living room. Computers invite a one-on-one interaction; it is not easy to sit around one with others. They are purely functional and people rarely develop an attachments to them.

But it could be otherwise. The designer Michael McCoy has shown that it is possible to design the parts of a computer with which you come into contact daily in a more "haptic and kinesthetic" way (McCoy 1997, 194–96). Thus one could use other materials and colors than the beige plastic of today, which would be one important step in taking the computer out of the office context. Sven Adolph, the designer of the space heater just discussed, has designed a pocket business computer with a built-in camera whose exterior is leather, making it look like a wallet. Caroline Nevejan and Marleen Stikker, of the Society for Old and New Media, won the Rotterdam Design Prize in 1997 for their "reading table for old and new media." It consisted of the familiar stem table of a cafe, covered with newspapers and magazines, but also had built-in computer screens and keyboards. People could come in, set coffee or beer mugs on the table, glance at the newspaper, and even check their email at it (Van Hinte 1997, 191–205). Such products mediate their environments in ways that turn them into something other than a business context and allow them to enter daily life.

I have just discussed two ways in which products can be designed to invite engagement. First, transparency can invite involvement when products become present-at-hand, that is, when they break or need to be revised. Second, products can also invite involvement from their readiness-to-hand. Engaging products, when they are used, can solicit human participation both in their functioning and in their suitability for integration in daily practices so that their users can make these products their own. There is also a third way in which products can invite involvement and attraction: their aging process may offer points of departure, as exemplified by the textile designs of Sigrid Smits. By the incorporation of seams and prints into furniture coverings that do not become visible until they age, for example, not only does it become possible for the pieces of furniture to effectively renovate themselves as they age rather than grow ugly, but people also become actively involved in that process. Furniture that is covered by Smits's textiles bears the traces of the history that they have had with their users. After all, they always age on the spots where people sit most often. Adolph's leather pocket computer is another example of a product that involves users in its aging. The knobs that operate the device are covered with leather, and in the course of time these grow steadily darker in color thanks to the oil on the human fingers that come into contact with it.

Conclusion

The material aesthetics that I have elaborated from the postphenomenological perspective points the way toward different design possibilities when it is applied to culturally durable product development. The most important viewpoint in this connection is the necessity of a materially oriented design approach. If products are to be designed to encourage human attachment, it is necessary to design them so that humans deal with the products *themselves* and not only with what they do or signify.

When only the functionality of products takes center stage, we are merely involved with what products *do* and not with *how* they do it; there is hardly any attention given to the material product that is present here and now. Additionally, design approaches that are exclusively oriented to the sign-character of products offer an inadequate grip on inviting an attachment with products. Objects to which people are attached on the basis of their meanings cannot be designed for in advance, for this kind of attachment arises out of the memories and associations that develop only over the course of years. And insofar as the sign-character of an object tries to tune in to the lifestyle of its users, this only amounts to a condition for possible attachment rather than an actual one. It may allow users to be open to the product, but in principle other products might also serve to exhibit the same lifestyle. Besides, lifestyles and the signs that express them are heavily influenced by fashion. What things do or mean for people is a role that can also be filled by other things—with the exception being those things to which people are strongly attached.

A material-aesthetic perspective allows the mediating role of products to be anticipated, with particular attention to their sensorial aspects. In connection with the ambition to create a durable relation between people and products, it allows the formulation of the design criteria of transparency and engaging capacity. Transparency promotes attachment to a product, for human relationships with it do not need to end when the product breaks. The engaging capacity of products invites attachment during the product's use by allowing trusted interaction with it and by involving people in the functioning and aging processes. In both cases a sensorial relation with objects as material artifacts arises, through which people are actually engaged with the very product that is present here and now. This engagement, supported by functionality and significance, amounts to a condition for a durable relation with these things.

The attachment to products that can arise by virtue of their transparency and engaging capacity is different from the emotional attraction particular things arouse in us thanks to the memories with which they are associated or their family references. The sought-for attraction does not come about from an involvement with the meanings of objects but from the practices they make possible—they are useful objects, not trophies or souvenirs. Precisely through such practices human beings develop a relationship with the materiality of the products concerned, rather than with their sign-character, even though that sign-character unmistakably plays a role in the relation. When people find that some product does not suit them they will not use it at all, but as soon as engaging and transparent products are used, they cannot get around their materiality as it presents itself in their functioning, wearing out, repair, and revision.

To be sure, it is not a foregone conclusion that transparent and engaging products will succeed in establishing a durable relationship with their users. A philosophical analysis of the different forms that the relations between people and products can assume cannot arrive at empirical conclusions about the relations that will actually arise. But this analysis does make clear that a durable relation with objects is unlikely to arise if people's engagement is not directed to the materiality of those products.

In this discussion of environmentally friendly product development I have not touched on all the details that the postphenomenological perspective is able to address. I have focused primarily on the mediation of action and engagement rather than on perception and experience. Moreover, I have strongly emphasized engagement with the mediating artifact itself and not with the context or with what it makes available. The above discussion is meant to be only one of the ways in which the postphenomenological perspective can be applied to concrete design practices; other explorations and applications are also possible. Information and communication technologies, for instance, mediate in increasing ways both human experience and existence. Here, too, the postphenomenological perspective offers a suitable beginning point from which to analyze the role that these technologies play in the relation between people and world, and to open the way for designers to anticipate this role.

In general, technological design processes would do well to try explicitly to anticipate the future mediating roles that products will play. These roles always exists, which gives designers the responsibility to carefully manage them. Because products by definition coshape the existence and experiences of people, their design is unavoidably a moral activity. Products help to provide answers to the question, "How should we live?" If designers fail to take account of this, they are neglecting an important dimension of their products.

CONCLUDING REMARKS

To the things themselves! I referred to this birth-cry of phenomenology in the introduction to this book. What, in the interim, has it turned up? First of all, it has become clear that an approach to technology in terms of concrete technological artifacts is essential in the philosophy of technology. A detailed analysis of the classical philosophy of technology revealed that the lack of such a perspective is at fault for its one-sided and simplified perspective on technology. Technology is reduced to its conditions of possibility, such as a technological way of thinking and a functional outfitting of society, that are required to keep our technological culture working. But the conclusions that this way of thinking and this social organization lead to loss of meaning and loss of self is much too premature. The pronouncements that the classical philosophy of technology made about technology pertained to the *conditions* of technology rather than to technology *itself*. I called this style of reasoning "transcendentalism," and showed how it approached technology in terms of its conditions of possibility, then absolutized its conclusions. It thought about technology in a "backward" fashion, from specific technologies to that from which they arose, and on the basis of which they exist, while in order to develop an adequate picture of technology it is also necessary to think "forward," from the specific technological artifacts with which human beings deal to the experiences and ways of existing that these technologies shape.

A suitable framework for thinking from the perspective of things can be provided by a reinterpretation of phenomenology in a way that can be called postphenomenological. In this perspective the relation between human beings and their world takes center stage, and are viewed as mutually constituting each other—human beings are what they are thanks to the ways in which they are present in their world, and their world is what it is thanks to how it appears to them. Things play a role precisely in this relation between human beings and world. This relation happens "via" things: human beings act with the help of artifacts and perceive through them. This role of things can be characterized as "mediation." Thanks to their mediating roles things help to shape the way in which human beings are involved with their world and interpret it. Things—and in our current culture especially technological artifacts—mediate how human beings are present in their world and how the world is present to them; they shape both subjectivity and objectivity.

In Part 2 I formulated a vocabulary with which to analyze this mediation. This vocabulary offers the possibility of describing technology in a more nuanced way than was possible within the language of the classical philosophy of technology. Technologies appear to be more than functional artifacts that presuppose a calculative way of thinking. To be sure, they reduce particular interpretations and forms of involvement, but also strengthen others and even create new forms of contact between human beings and their world.

In this final chapter of the book I made it clear that a postphenomenological perspective provides a more elaborate fleshing out of the aesthetics of products than was possible from the semiotic perspective. I did so on the basis of the original meaning of the word "aesthetics": the study of the senses. By approaching products only as signs the emphasis falls too one-sidedly on visual aspects, while from the perspective of their uses all of the other senses also become important. Giving due attention to these sensorial dimensions of the use of objects will make their materiality again relevant. The sensorial dimension forms a tangent plane between human beings and world in a way that brings this material aesthetics in direct connection with mediation. When technological artifacts are used, they always help to shape the relation between human beings and their world in specific ways. Things codetermine how human beings experience their world and how their existence unfolds in it. A material aesthetics makes it possible to anticipate this mediating ability of products and incorporate it into the design process.

The design of technology thereby becomes no longer an internal technological affair, but appears to be a moral matter as well. Technologies are not merely functional objects that also have dimensions of style and meaning; they mediate the relations between human beings and their world, and thereby shape human experiences and existence. Technologies help to determine how people act, so that it is not only people but also things who give answers to the classical moral question, "How to live?" It is time that we take the contributions of technology seriously and combine our forces to provide new answers to this ancient question that still applies to the technological world in which we live.

Bibliography

Achterhuis, H., ed. 1992. *De maat van de techniek.* Baarn: Ambo.

———, ed. 1993. *Deugt de ethiek?* Baarn: Gooi en Sticht.

———. 1993a. "Het permanente gemoraliseer." In *Deugt de ethiek?* ed. H. Achterhuis. Baarn: Gooi en Sticht.

———. 1995. "De moralisering van de apparaten." *Socialisme en Democratie* 52 (1): 3–12.

———. 1998. *De erfenis van de utopie.* Amsterdam: Ambo.

———, ed. 2001. *American Philosophy of Technology: The Empirical Turn.* Trans. R. Crease. Bloomington: Indiana University Press.

———. 2001a. "Introduction: American Philosophers of Technology." In *American Philosophy of Technology,* ed. H. Achterhuis and trans. R. Crease. Bloomington: Indiana University Press.

Akrich, M. 1992. "The De-scription of Technical Objects." In *Shaping Technology/Building Society,* ed. W. E. Bijker and J. Law. Cambridge, Mass.: MIT Press.

Anders, G. 1987. *Die Antiquiertheit des Menschen.* Munich: C. H. Beck.

Baird, G. 1976. "The Dining Position: A Question of 'Langue' and 'Parole.'" *Dutch Forum,* October.

Banham, R. 1981. *Design by Choice: Ideas in Architecture.* Ed. Penny Sparke. London: Academy Editions.

Baudet, H. 1986. *Een vertrouwde wereld: 100 Jaar innovatie in Nederland.* Amsterdam: Bert Bakker.

Belt, H. van den, and J. Keulartz. 1996. "De technische en de palliatieve rede." *Krisis* 64:85–91.

Benjamin, W. 1968. *Illuminations.* New York: Harcourt, Brace, and World.

Berg, J. H. Van den. 1970. *Things: Four Metabletic Reflections.* Pittsburgh: Duquesne University Press.

Biemel, W. 1989. *Martin Heidegger.* Reinbek bei Hamburg: Rowohlt Taschenbuch Verlag.

Bijker, W. E. 1995. *Of Bicycles, Bakelites, and Bulbs: Toward a Theory of Sociotechnical Change.* Cambridge, Mass.: MIT Press.

Bijker, W. E., and J. Law, eds. 1992. *Shaping Technology/Building Society*. Cambridge, Mass.: MIT Press.

Bolle, E. 1985. *Afscheid van wat nooit geweest is.* Groningen: Konstapel.

Bollnow, O. F. 1955. *Existenzphilosophie.* Stuttgart: W. Kohlhammer Verlag.

Borgmann, A. 1984. *Technology and the Character of Contemporary Life.* Chicago: University of Chicago Press.

———. 1987. "The Question of Heidegger and Technology." *Philosophy Today* (Summer): 98–162.

———. 1992a. *Crossing the Postmodern Divide.* Chicago: University of Chicago Press.

———. 1992b. "'Cosmopolitanism and Provincialism: On Heidegger's Errors and Insights." *Philosophy Today* (Summer): 131–45.

———. 1995. "The Depth of Design." In *Discovering Design: Explorations in Design Studies,* ed. R. Buchanan and V. Margolin. Chicago: University of Chicago Press.

Bossche, M. van den. 1995. *Kritiek van de technische rede: Een onderzoek naar de invloed van de techniek op ons denken.* Utrecht: Jan van Arkel; Leuven: Van Halewyck.

Buchanan, R., and V. Margolin, eds. 1995. *Discovering Design: Explorations in Design Studies.* Chicago: University of Chicago Press.

Bürdek, B. 1996. *Design: Geschiedenis, theorie en praktijk van de produktontwikkeling.* The Hague: Ten Hagen & Stam.

Cockburn, C., and S. Ormrod. 1993. *Gender and Technology in the Making.* London: Sage.

Coolen, M. 1992. *De machine voorbij.* Meppel en Amsterdam: Boom.

Csikszentmihalyi, M., and E. Rochberg-Halton. 1995. *The Meaning of Things: Domestic Symbols and the Self.* Cambridge: Cambridge University Press.

Dijk, P. van. 2000. *Anthropology in the Age of Technology: The Philosophical Contribution of Günther Anders.* Value Inquiry Book Series 103. Amsterdam: Rodopi.

Diop, B. 1985. "Breath." In *Poems of Black Africa,* ed. Wole Soyinka. London, Nairobi, Ibadan: Heinemann.

Dreyfus, H. 1992. "Heidegger's History of the Being of Equipment." In *Heidegger: A Critical Reader,* ed. H. Dreyfus and H. Hall. Oxford: Blackwell.

Dreyfus, H., and H. Hall, eds. 1992. *Heidegger: A Critical Reader.* Oxford: Blackwell.

Eternally Yours. 1997. *Lectures Given During the Eternally Yours Congress, 1997.* The Hague: Eternally Yours.

Feenberg, A. 1999. *Questioning Technology.* London: Routledge.

———. 2004. "From Essentialism to Constructivism: Philosophy of Technology at the Crossroads." http://www-rohan.sdsu.edu/faculty/feenberg/talk4.html (accessed 8 March 2004).

Forty, A. 1992. *Objects of Desire.* London: Thames & Hudson.

Fry, T. 1994. *Remakings: Ecology/Design/Philosophy.* Sydney: Envirobooks.

Heidegger, M. 1966. "The Memorial Address." In *Discourse on Thinking,* trans. J. M. Anderson and E. H. Freund. New York: Harper & Row.

———. 1971a. "The Origin of the Work of Art." In *Poetry, Language, Thought,* trans. A. Hofstadter. New York: Harper & Row.

———. 1971b. "Building, Dwelling, Thinking." In *Poetry, Language, Thought,* trans. A. Hofstadter. New York: Harper & Row.

———. 1971c. "The Thing." In *Poetry, Language, Thought,* trans. A. Hofstadter. New York: Harper & Row.

———. 1977a. "The Question Concerning Technology." In *The Question Concerning Technology and Other Essays,* trans. W. Lovitt. New York: Harper & Row.

———. 1977b. "The Turning." In *The Question Concerning Technology and Other Essays*, trans. W. Lovitt. New York: Harper & Row.

———. 1977c. "The Age of the World Picture." In *The Question Concerning Technology and Other Essays*, trans. W. Lovitt. New York: Harper & Row.

———. 1993. "What Calls for Thinking?" In *Basic Writings*, trans. D. F. Krell. New York: Harper & Row.

———. 1996. *Being and Time*. Trans. J. Macquarrie and E. Robinson. New York: Harper & Row.

Heij, P. 1995. "J. H. van den Berg: Eerst en vooral fenomenoloog." In *Metabletische perspectieven: Beschouwingen rond het werk van J. H. van den Berg*, ed. W. Vandereycken and J. de Visscher. Leuven and Amersfoort: Acco.

Heskett, J. 1980. *Industrial Design*. London: Thames & Hudson.

Hinte, E. 1996a. "Binnen zonder kloppen." In *Jan Bart Klasterprijs voor de kunstkritiek*. Amsterdam: Sichting Jan Bart Klasterprijs voor de kunstkritiek. [Occasional edition; also published in *Items* 2, 1995].

———. 1996b. "Lang zullen ze leven." In *Jan Bart Klasterprijs voor de kunstkritiek*. Amsterdam: Sichting Jan Bart Klasterprijs voor de kunstkritiek. [Occasional edition; also published in *Items* 8, 1995].

———, ed. 1997. *Eternally Yours: Visions on Product Endurance*. Rotterdam: 010 Publishers.

Hottois, G. 1996. *Symbool en techniek*. Kampen: Kok Agora; Kapellen: Pelckmans.

———. 1996a. "Staat de technowetenschap buiten de filosofie?" In *Symbool en techniek*. Kampen: Kok Agora; Kapellen: Pelckmans.

Husserl, E. 1970. *Logical Investigations*. Trans. J. N. Findlay. 2 vols. London: Routledge & Kegan Paul.

Hybašek, E. 1984. *Das Menschenbild bei Karl Jaspers*. Graz: Dissertationen der Karl-Franzens-Universität.

Ihde, D. 1976. *Listening and Voice*. Athens: Ohio University Press.

———. 1979. *Technics and Praxis*. Dordrecht: Reidel.

———. 1983. *Existential Technics*. Albany: SUNY Press.

———. 1990. *Technology and the Lifeworld*. The Indiana Series in the Philosophy of Technology. Bloomington: Indiana University Press.

———. 1991. *Instrumental Realism*. The Indiana Series in the Philosophy of Technology. Bloomington: Indiana University Press.

———. 1993a. *Philosophy of Technology: An Introduction*. New York: Paragon House.

———. 1993b. *Postphenomenology*. Evanston: Northwestern University Press.

———. 1998. *Expanding Hermeneutics*. Evanston: Northwestern University Press.

Jaspers, K. 1951. *Man in the Modern Age*. Trans. E. Paul and C. Paul. London: Routledge & Kegan Paul.

———. 1953. *The Origin and Goal of History*. Trans. M. Bullock. New Haven: Yale University Press; London: Routledge & Kegan Paul. Reprint, Westport, Conn.: Greenwood Press, 1976.

———. 1963. *The Atom Bomb and the Future of Man*. Chicago: University of Chicago Press.

———. 1969–71. *Philosophy*. Trans. E. B. Ashton. 3 vols. Chicago: University of Chicago Press.

Keulartz, J. 1998. "The French Connection: Inleiding tot het Heidegger-debat." *K&M—Tijdschrift voor empirische filosofie* 2:155–68.

Kierkegaard, S. 1944. *Concluding Unscientific Postscript*. Trans. D. F. Swenson. Princeton: Princeton University Press.

Kockelkoren, P. 1992. *De natuur van de goede verstaander*. Proefschrift, Enschede: Universiteit Twente.

———. 1994. "Naar een technische intimiteit met de dingen." *Wijsgerig Perspectief* 34 (6): 187–93.

———. 1996. "De esthetische wending in het natuurdebat." *Bzzletin* 240 (November): 49–67.

"Kortere steel tegen luieren." *Algemeen Dagblad* (The Netherlands), March 23, 1996.

Latour, B. 1992. "Where Are the Missing Masses? The Sociology of a Few Mundane Artifacts." In *Shaping Technology/Building Society*, ed. W. E. Bijker and J. Law. Cambridge, Mass.: MIT Press.

———. 1993. *We Have Never Been Modern*. Trans. C. Porter. Cambridge, Mass.: Harvard University Press.

———. 1994. "On Technical Mediation: Philosophy, Sociology, Genealogy." *Common Knowledge* 3:29–64.

———. 1997. *De Berlijnse sleutel*. Amsterdam: Van Gennep.

———. 1999a. *Pandora's Hope*. Cambridge, Mass.: Harvard University Press.

———. 1999b. "On Recalling ANT." In *Actor-Network Theory and After*, ed. J. Law and J. Hassard. Oxford: Blackwell.

Law, J., and J. Hassard, eds. 1999. *Actor-Network Theory and After*. Oxford: Blackwell.

Manzini, E. 1995. "Prometheus of the Everyday: The Ecology of the Artificial and the Designer's Responsibility." In *Discovering Design: Explorations in Design Studies*, ed. R. Buchanan and V. Margolin. Chicago: University of Chicago Press.

McCoy, M. 1997. "Angling for Endurance." In *Eternally Yours: Visions on Product Endurance*, ed. E. van Hinte. Rotterdam: 010 Publishers.

McDermott, C. 1992. *Essential Design*. London: Bloomsbury.

Merleau-Ponty, M. 1962. *Phenomenology of Perception*. Trans. Colin Smith. London: Routledge & Kegan Paul.

Meyer, G. De. 1994. *De zin van de onzin: De cultuur van de slechte smaak*. Antwerpen and Baarn: Hadewijch.

Mitcham, C. 1994. *Thinking Through Technology: The Path Between Engineering and Philosophy*. Chicago: University of Chicago Press.

Mol, A. 1997. *Wat is kiezen? Een empirisch-filosofische verkenning*. Oratie, Enschede: Universiteit Twente.

Mol, A., and J. Mesman. 1996. "Neonatal Food and the Politics of Theory: Some Questions of Method." *Social Studies of Science* 26:419–44.

Muller, W. 2001. *Order and Meaning in Design*. Utrecht: Lemma.

Ortega y Gasset, J. 1972. "Thoughts on Technology." In *Philosophy and Technology*, ed. C. Mitcham and R. Mackey. New York: Free Press.

Pöggeler, O. 1991. *Martin Heidegger's Path of Thinking*. Atlantic Highlands, N.J.: Humanities Press.

Popkema, M., et al. 1997. "Technologie en Zwangerschap: De politiek van een prenatale screeningstest." *K&M—Tijdschrift voor empirische filosofie* 2:97–123.

Procee, H. 1997. *De Nieuwe Ingenieur*. Amsterdam: Boom.

Ramakers, R. 1986. "Van Morris tot Memphis." In *Vormgeving in veelvoud: Over vormen, beelden en beeldvorming*, ed. R. van Rossum et al. Delft: Eburon.

Rapp, F. 1994. *Die Dynamik der modernen Welt: Eine Einführung in die Technikphilosophie*. Hamburg: Junius Verlag.

Richardson, W. J. 1963. *Heidegger: Through Phenomenology to Thought*. The Hague: Martinus Nijhoff.

Riele, H. te, and A. Zweers. 1994. *Eco-design: Acht voorbeelden van milieugerichte produktontwikkeling.* Delft: TNO Produktcentrum; The Hague: NOTA.

Rip, A. 1978. *Wetenschap als mensenwerk.* Baarn: Ambo.

———. 1988. *Technologie als mensenwerk.* Oratie, Enschede: Universiteit Twente.

Robinson, J. 1977. *How Americans Use Time: A Social-Psychological Analysis of Everyday Behavior.* New York: Praeger.

Rossum, R. van, et al. 1986. *Vormgeving in veelvoud: Over vormen, beelden en beeldvorming.* Delft: Eburon.

Saner, H. 1970. *Karl Jaspers.* Reinbek bei Hamburg: Rowohlt Taschenbuch Verlag GbmH (Rororo Bildmonographieen).

Schneiders, W. 1965. *Karl Jaspers in der Kritik.* Bonn: H. Bouvier.

Schuurman, E. 1980. *Technology and the Future: A Philosophical Challenge.* Toronto: Wedge Publishing.

Schwartz, M., and R. Jansma. 1988. *De technologische cultuur.* Amsterdam: De Balie.

Seubold, G. 1986. *Heideggers Analyse der neuzeitlichen Technik.* Freiburg and Munich: Verlag Karl Alber.

Smits, M. 2001. "Langdon Winner: Technology as a Shadow Constitution." In *American Philosophy of Technology,* ed. H. Achterhuis and trans. R. Crease. Bloomington: Indiana University Press.

Soyinka, W., ed. 1985. *Poems of Black Africa.* London, Nairobi, Ibadan: Heinemann.

Sparke, P. 1986. *An Introduction to Design and Culture in the Twentieth Century.* London: Allen & Unwin.

Springer, J. L. 1951. *Existentiële metaphysica: Inleiding in de metaphysica van Karl Jaspers.* Assen: Van Gorcum.

Swierstra, T. 1999. "Moeten artefacten moreel gerehabiliteerd?" *K&M—Tijdschrift voor empirische filosofie* 4:317–26.

Tijmes, P. 1991. "De werkelijkheid in eerste en tweede structuur." *De uil van Minerva* 8 (1): 31–38.

———. 1992a. "Jacques Ellul: Autonome techniek." In *De maat van de techniek,* ed. H. Achterhuis. Baarn: Ambo.

———. 1992b. "Heidegger: Techniek als metafysica." In *De maat van de techniek,* ed. H. Achterhuis. Baarn: Ambo.

———. 2001. "Albert Borgmann: Technology and the Character of Everyday Life." In *American Philosophy of Technology,* ed. H. Achterhuis and trans. R. Crease. Bloomington: Indiana University Press.

Vandereycken, W., and J. de Visscher, eds. 1995. *Metabletische perspectieven: Beschouwingen rond het werk van J. H. van den Berg.* Leuven and Amersfoort: Acco.

Verbeek, P. P. 1996. "De dingen doorzien: Het technologisch ontsluiten van de natuur." *Filosofie en Praktijk* 17 (3): 113–26.

———. 1999. "De daadkracht der dingen: Esthetica na de dood van het object." *De Gids* 6:401–13.

———. 2000. "The Thing About Technology." *Research in Philosophy and Technology* 19:281–99.

———. 2001. "Don Ihde: The Technological Lifeworld." In *American Philosophy of Technology,* ed. H. Achterhuis and trans. R. Crease. Bloomington: Indiana University Press.

———. 2002. "Devices of the Good Life: On Borgmann's Philosophy of Information and Technology." *Techné* 6 (1): 69–92.

Verbeek, P. P., and P. Kockelkoren. 1998. "The Things That Matter." *Design Issues* 14 (3): 28–42.

Vries, G. de. 1999. *Zeppelins: Over filosofie, technologie en cultuur.* Amsterdam: Van Gennep.

Wal, K. van der. 1970. *Karl Jaspers, filosoof van de vrijheid.* Baarn: Wereldvenster.

Winner, L. 1986. *The Whale and the Reactor.* Chicago: University of Chicago Press.

———. 1986a. "Do Artifacts Have Politics?" In *The Whale and the Reactor.* Chicago: University of Chicago Press.

Woolgar, S., and G. Cooper. 1999. "Do Artefacts Have Ambivalence?" *Social Studies of Science* 29 (3): 433–49.

Zaccai, G. 1995. "Art and Technology: Aesthetics Redefined." In *Discovering Design: Explorations in Design Studies,* ed. R. Buchanan and V. Margolin. Chicago: University of Chicago Press.

Zimmerman, M. 1990. *Heidegger's Confrontation with Modernity: Technology, Politics, Art.* Bloomington: Indiana University Press.

Index

abortion, 44
abstractness (of classical philosophy of technology), 4
abstractness, 60, 65
Achterhuis, H., 16 n. 1, 36 n. 8, 117, 213, 216
actant, 149
action, program of, 156–61, 156 n. 5
actor-network theory, 102, 112, 149–52
  versus postphenomenology, 161–68
Adolph, S., 230, 232
aesthesis, 211
aesthetics
  and ethics, 212–17
  and the senses, 210–12
  material, 12, 204, 209, 218, 230, 232–33
agricultural techniques, 61–62
ahistorical versus historical approach of technology, 72–75
AIO-data, 205–6
alètheia, alètheuein, 50, 85, 87
alienation, 5, 7, 16, 22–23, 106, 113
alterity relation, 123, 126–27, 195
ambivalence of technology, 135, 191, 193
amodernism, 152–54
amplification
  of engagement, 190–95
  of perception, 131–35
Anders, G., 16, 27, 28, 29
anthropological view of technology, 49
Apparatus, the, 18, 20, 22
Aristotle, 52
art deco, 25
art nouveau, 24
art, artwork, 28, 29, 57, 58, 83–88, 210
  products as, 211
Arts and Crafts Movement, 24
assembly line, 35, 99
Asterix, 154
astronomy, 142
atomic bomb, 42, 61–62
attachment (between humans and things), 12, 20–23, 218, 223–25, 232–33
authenticity
  of existence 16, 20–22, 36
  of things, 2, 27
automobile. *See* car
autonomy, 33
  of human beings, 33
  of technology, 101, 136
availability, 176

Bach, J. S., 188, 190
background of technology, 177
background relation, 123, 127–28
backward (thinking), 7–9, 29, 44, 67, 88
Bacon, F., 176
Baird, G., 207
Banham, R., 26
Baudet, H., 115, 189, 217
beauty, 40, 85
being, 51, 55
  as coming into being, 56, 73–75
  historicity of, 51, 61, 70–75, 92
  in Jaspers's work, 34
Benjamin, W., 27–29
Berg, J. H van den, 105–7, 132–34
Bestand, bestellen, 54–55, 81
bicycle, 102
Bijker, W., 102, 217
binoculars, 132–33
black box, black boxing, 103, 158–59, 166

Blonk, 220 n. 6
blueprint, 27
Blut und Boden, 59
Bolle, E., 73
bond (between humans and things). *See* attachment
Boorstin, D., 181
Borgmann, A., 3, 48 n. 2, 136 n. 4, 173–99
boundary situations, 33, 36
bridge, 68, 70, 71, 91
Bürdek, B., 206
bureaucracy, 18, 36

car, 43, 149, 212
care, 110
  ethics of, 215
  for objects, 215, 221
Carr, D., 228
cassette recorder, 114 n. 10
causality, 52–53
CD player, 188–90, 194
choice, 33, 36
classical philosophy of technology, 4–8, 27, 174
Cluj, 115
Cockburn, C., 5, 190
commodity, 19, 177, 227
communication technology, 66, 137, 233
composition, 156–58
concealed, the, 50
conditioning, 212
conditions of possibility, 6–7, 102
  in Heidegger's approach, 65, 71–72
  in Jaspers's approach, 23, 29–30, 36, 38, 44
connotative functions. *See* functions
consciousness, 106, 112
consequentialism, 214–15
conservatism, 60
constructivism, 102–4, 115, 142, 150
  social. *See* social constructivism
consumption, 178, 188
contemporary style, 25
contrast, 131, 134
Coolen, M., 7, 7–8 n. 3
Cooper, G., 116 n. 13
copy, 28
Corbusier, Le, 25
couch potato, 179
Csikszentmihalyi, M., 223–25
culture, 129, 135–38

Dasein
  in Heidegger's work, 77
  in Jaspers's work, 32, 34
deictic discourse, 184–85
delegation, 159–61, 169–70
democracy, liberal, 180–83, 192
demonism, 17, 22, 38, 39
denotative functions. *See* functions
deontology, 214–15
dependence on technology, 18, 35
Descartes, R., 52, 176
design, industrial, 24, 203–36
determinism, 5, 101
device (Borgmann), 177
device paradigm, 175, 178
Dijk, P. van, 27
Diop, B., vii
door-spring, 160
Down's syndrome, 33
Dreyfus, H., 81, 81 n. 17
dwelling, 59

earth versus world (Heidegger), 86–87
effort, 186–87
Ellul, J., 15, 174
eloquence of reality, 187
embodiment, 123–26
empirical approach of technology, 4, 6, 9, 100–104
engagement, 179–80, 182, 184, 228
  mediation of, 191–95
  two forms of, 186–88
  with devices, 188–91, 194, 228–32
enframing. See *Gestell*
engaging capacity, 233
Enlightenment, 148, 152, 176
ens summum, 52
environment, 24
epistèmè, 139
epochè, 109
equality, 181
equipment. *See* tools
ereignis, 74
essence, essentialism, 113, 149–50, 156

essential intuition. See *Wesensschau*
Eternally Yours, 12, 24, 27, 218–23
ethics, 212–17
  and artifacts, 212–17
  of behavior, 212
  virtue, 212
ethnomethodology, 151, 162
existence, 17, 23, 30, 38, 149
  in actor-network theory, 156, 164
existential philosophy, 16, 31–34, 110–11
existentialism, 149
experience, 112, 122
eyeglasses, 130–31, 133

fashion, 232
Feenberg, A., 61, 61 n. 6, 62, 65
flow, 224
focal practices, 184
focal things, 184
form follows fun, 26
form follows function, 25
Forty, A., 209
forward (thinking), 8, 29, 38, 67, 88
Foucault, M., 139
foundations, 113
fourfold, 69, 72, 74, 89, 173 n. 1
Frankfurt School, 61
freedom, 32, 34, 181
Fry, T., 221 n. 7
function, functionality, 19, 25, 35, 156, 205, 208–10
  connotative, 205–6
  denotative, 205–6
  practical, 205–6
  primary, 206
  product language, 205–6
  secondary, 206–8
functional clarity, 227
functionalism, 8, 20, 30, 36

Gehlen, A., 36 n. 8
Gelassenheit, 58, 67, 94
Geschick, 55–56, 65
Gestell, 55–57, 61, 64–66, 158 n. 7, 173 n. 1
Geviert. *See* fourfold
God, 2, 52, 56
Gogh, V. van, 85
good life, 37, 178–85, 192, 212
Gros, J., 205
guiding technology, 41–42
Guignon, Ch., 73 n. 12
gun, 154–55
Günther, G., 2

hammer, 78, 123
handiness. *See* readiness-to-hand
heating, 128, 177, 230
Hegel, G. W. F., 31, 163
Heidegger, M., 27, 47–95, 210, 225
  and Borgmann, 173, 194
  and classical philosophy of technology, 3, 7 n. 3, 10
  and hermeneutic phenomenology, 108, 111, 134
  and technological mediation, 114, 124
  externalist critique of, 60
  on science, 140
  versus Latour, 155 n. 4, 158 n. 7
Heij, P., 105 n. 4
hermeneutic relation, 126
hermeneutics, 111, 119, 126
  of things, 141
  technology and, 128–35
Heskett, J., 205 n. 1, 209
Hinte, E. van, 222, 224, 226–27, 231
Hölderlin, F., 57
Holocaust, 61
Hottois, G., 9, 66–67
Hughes, J. D., 69
human-artifact relations, 122–35
Husserl, E., 12, 77–78, 105 n. 5, 107–10, 129 n. 3
Hybašek, E., 21 n. 3, 38
hybrid, 152–54
hydroelectric plant, 62–65, 68–70, 72–75, 81, 92

ideal type, 19, 26
idealism, 111, 143
identity, 32
Ihde, D., 121–45, 164 n. 9
  and postphenomenology, 108, 109 nn. 7–8, 112–18
  on Heidegger, 48 n. 2, 68–71, 78
imaging technology, 137

implication (Borgmann), 179
individuality, 35
industrial design. *See* design, industrial
industrial production, 81–82
Industrial Revolution, 17
industrialization, 99
information and communication technologies, 36, 137, 234
infrared photography, 134
inhibition, 171
inscription, 159–61
instrumental view of technology, 49
instrumentalism, 11, 136, 174
intellect, 41
intention, 156, 216
intentionality, 78, 109, 116, 122, 129 n. 3
  cultural, 137
  existential, 78 n. 14
  instrumental, 123
  technological, 114–16, 137
interaction (with technology), 127
interpretation, 125–26
  technology and, 128–35
interpretive flexibility, 217
invitation, 171
involvement, three variants of, 192
irony of technology, 179–80

Jaspers, K., 3, 10, 15–46, 58, 180, 185
  and existential phenomenology, 111
Jobs, S., 227
jug, 89–91
Jugendstil, 24

Kant, I., 7, 163
Kehre, 49, 73, 77, 83
key ring, 157
Kierkegaard, S., 31, 32, 110
Kockelkoren, P., 218 n. 5
Kuhn, Th., 139

labor, 18, 43
language games, 113
Latour, B., 102–4, 141, 147–72, 207
leisure, 35, 43
lifespan of products, 220
lifestyle, 2, 26, 207, 232
lifetime, psychological, 12
limits of technology, 39–41
linguistic turn, 1, 104–5, 112
longevity, 219

machinery, 177, 227
  rehabilitation of, 229
macroperception, 123
  in science, 139–40
  mediation of, 129, 131–35
Manzini, E., 12, 220–21, 224
Marx, K., 163
mass consumption, 180
mass culture, 15
mass existence, 31–38
mass production, 15, 23, 28, 30, 180
mass products, 29
mass rule, 17–18, 21–23
mass, 21–22
materialism, 2, 27
materiality, 5, 8, 208, 225, 229
McCoy, M., 231
McDermott, C., 24
meaning of products, 5, 205
meaningfulness, 186
means (technology as), 39
mechanization, 18, 25, 35
mediation, 11, 114–16, 128–45, 154–61, 168–71
  and functionality, 208
  anticipating, 217–18
  relation of, 123–26
Merleau-Ponty, M., 106–8, 123, 162
Mesman, J., 150 n. 2
metabletics, 132
metaphysics, 64 n. 7
MET-approach, 219
Meyer, G. de, 189
microperception, 123
  in science, 139–43
  mediation of, 128, 130–31
microscope, 133–34
microwave oven, 11, 102, 189
Middle Ages, 51, 56
mill, 72, 74, 75
Mitcham, C., 16 n. 1, 100
modernism, 24–25, 204–5, 208
modernity, 152
Mol, A., 32, 150 n. 2

morality of artifacts, 212, 214–16
Morris, W., 24
Moses, R., 116–17, 216
MRI, 66, 142
Muller, W., 204–7, 217
multiculturalism, 137
multistability, 117–18, 136–37, 170, 207, 217
music, 188–89, 192, 194, 224

natural attitude, 109
navigation, 136
Nazi party, 59
Necker cube, 118
neediness, 37, 38
neutrality (of technology), 39, 43, 136
Nevejan, C., 231
nostalgia, 60, 67, 71

objectivity, 31, 112, 135, 142, 171. *See also* subject, subjectivity
obsolescence, 27
ontic versus ontological, 61, 63–66
ontological difference, 51
operativity of technology, 9
operativity, 67
original, 29
Ormrod, S., 5, 190
Orpheus, Orphic, 6–8, 113
Ortega y Gasset, J., 37, 37–38 nn. 9–10
ozone hole, 153–54

paradigm
  in Kuhnian sense, 139
  technology as, 174–78
Parthenon, 69
PDA (personal digital assistant), 197–99
pen, 115
perception, 22, 40,106, 112, 211. *See also* microperception, macroperception
  transformation of, 130–31
phenomenology, 104–11, 162–63, 167
  as method, 106–8
  existential, 10, 111, 119
  hermeneutical, 10, 111, 119
physis, 82
piano, 188, 194, 229–30
Plato, platonism, 1, 2, 27, 28, 51, 73
pluralism, 113, 175, 182
pluriculturalism, 137
Pöggeler, O., 55, 75, 86–87
poièsis, 52, 55, 57
politics of artifacts, 69
pop design, 26
Popkema, M., 33, 216
population growth, 18
positivism, 104
possibility, conditions of. *See* conditions of possibility
postmodernism
  in industrial design, 2, 26, 204–5
  in philosophy, 2, 104–5, 112, 153
postphenomenology, 10, 112–13, 163–64
  versus actor-network theory, 161–68
  versus semiotics, 207
praxis, 111
prenatal diagnostics, 33, 41
prescription, 160
present-at-hand, 79–80, 85, 194–95, 225–26, 229
Presocratics, 51
probe, 125
Procee, H., 115
program of action. *See* action
progress of technology, 41
promise of technology, 176
purification (of humans and nonhumans), 149

racism, 216
radio telescope, 140, 142
radio, 59
Ramakers, R., 205 n. 1
Rapp, F., 16 n. 1
reading table for old and new media, 231
ready-to-hand, 79–80, 85, 114, 124, 194–95, 225–29
realism, 112–13, 141, 143, 151
  instrumental, 141
reason, 41, 42
recycling, 220
reduction
  of engagement, 190–95
  of perception,131–35
  phenomenological. See *epochè*
reflection, essential, 57, 59
reflection, permanent, 36

reform of technology, 184–85
refrigerator, 128
relativism, 113, 122
releasement. See *Gelassenheit*
reliability, 85
repair, 220
representation
  art as, 85
  provided by technology, 126
res cogitans, 52
res extensa, 52
responsibility, 42
Richardson, W., 77 n. 13, 81 n. 16
Riele, H. te, 218–19
rights, moral, 215
Robinson, J., 179 n. 3
Rochberg-Halton, E., 223–25
romanticism, 68
running, 184

Sartre, J. P., 149
Schuurman, E., 7 n. 3
science, 105, 129, 139–43, 151
SCOT approach, 102
script, 115, 160, 207
Sein zum Tode, 27
self-realization, 181
semiotics, 150, 206–7
sending. See *Geschick*
senses, 209
Seubold, G., 63–64, 64 n. 7
shoes, 85
signs, products as, 2, 206, 210
Smits, M., 116 n. 12
Smits, S., 222, 232
social constructivism, 101–2, 150
Socrates, 213, 215
Sorge. *See* care
sovereignty, 43
Sparke, P., 205 n. 1
spectrogram, 131, 134
speed bump, 159, 209, 216
speed control, 213, 216
standard picture of technology, 3
standing-reserve, 52, 56, 63, 69, 82
Stikker, M., 231
streamlining, 25
structure, primary and secondary, 133–34
styling, 25
subject
  constitution of subject and object, 112, 116, 129–30, 171
  death of the, 2
  subject-object dichotomy, 117, 129, 148, 152–54
subjectivity, 31, 112
substantivism, 11, 136, 174
superfluousness, 37
sustainability, cultural, 220, 224–25
Swierstra, Tsj., 214–17
symmetry (actor-network theory), 102, 149–50, 169–71

table, 207
  culture of the, 184
techné, 52–54, 57, 72, 73, 81–82, 87
technocracy, 213, 216
technology
  as "enframing," 54–56, 62, 80
  as "revealing," 50–54, 80
  as interpretation of reality, 66
  as object, 67, 76, 80, 93
  classical philosophy of. *See* classical philosophy of technology
  definition of, 3, 100–101 n. 1
  empirical philosophy of. *See* empirical philosophy of technology
  essence of, 49, 57, 64 n. 7, 67, 93, 117–18
  human-technology relations, 122–35
  traditional technique, 67, 69, 93
technophobia, 9
telephone, 44, 114, 131
telescope, 131, 133
television, 59, 179, 187, 189
telos, technological, 114 n. 10
temple, 68, 86
thermometer, 126, 131
things
  in late Heidegger's sense, 70
  mere thing (Heidegger), 84
  the thing things, 47–48, 89–91
  turn toward, 3, 9
Tijmes, P., 16, 174 n. 2
tomography, 140
tools, 78–79, 83–88, 114, 123
totalitarianism, 213–14

touch, 211
train, 114
transactions between humans and things, 223
transcendence, 34, 53
transcendental fix, 100
transcendental, 7, 164
transcendentalism, 7–8, 11, 113
  in Heidegger's work, 65, 71, 88
  in Jaspers's work, 23, 27, 28, 44
transformation (of perception), 130–31
translation, 142, 155–56, 170–71
transparency, 126, 225–28, 233
truth, 50–51, 103, 113
turn toward things, 76
turn, linguistic, 1–2
typewriter, 115, 117

ultrasound, 140, 132
upgrading, 220
utility
  material, 204–6
  social-cultural, 204–8

Verbeek, P. P., 48 n. 3, 121 n. 1, 174 n. 2, 210 n. 3, 218 n. 5
Vernunft, 41
Verstand, 41
visualism, 210
vocabulary, postphenomenological, 195–96
Vorhandenheit. *See* presence-at-hand
Vries, G. de, 212–13
Vuitton, L., 2

Wesensschau, 105 n. 5
wheelchair, 130
will to power, 52, 66
windmill, 68
Winfrey, O., 6
Winner, L., 69, 116–17
Woolgar, S., 116 n. 13
word processor, 115, 188
Wozniack, S., 227

X-ray, 142

Zaccai, G., 210
Zimmerman, 48 n. 2, 52 n. 4, 73 n. 12
zuhandenheit. *See* readiness-to-hand
Zweers, A., 218–19